SURVIVAL

A HOSTILE WORLD

Written and illustrated by

BRIAN CAPON

TIMBER PRESS, Inc.
Portland, Oregon

Many thanks to
Eve Goodman of Timber Press
for her enthusiasm and guidance
during the preparation of the manuscript

TIMBER PRESS, Inc.
The Haseltine Building
133 S.W. Second Avenue, Suite 450
Portland, Oregon 97204-3527
U.S.A.

Printed in Singapore

Library of Congress Cataloging-in-Publication Data

Capon, Brian
    Plant Survival : adapting to a hostile world / written and illustrated
by Brian Capon.
        p. cm.
    Includes index.
    ISBN 0-88192-283-8
    1. Plants—Adaptation—Juvenile literature. [1: Plants. 2. Adaptation
(Biology) 3. Ecology.] I. Title.
QK912.C36 1994
581.5—dc20   93-43342
                                                                   CIP
                                                                   AC

PLANT SURVIVAL

PLANT
ADAPTING TO

CONTENTS

BRISTLECONE PINE

INTRODUCTION

THE WORST PART of being a plant is that if you don't like the place where you were born, you can't pick up and move. With roots firmly anchored to the ground, a plant is stuck in one location for life, which may be less than a year but could be for centuries. Some bristlecone pine and giant sequoia trees are still standing where they began to grow, three to four thousand years ago.

Compared with animals, plants must be much more ready to cope with constant changes in their environment if they are to survive. Animals can escape to shelter themselves from strong winds, heavy rains, cold winter nights, or the heat of the summer sun. If other creatures threaten to attack, animals can run, swim, or fly to safety.

Not so with plants. They must have ways to prevent destruction by hungry insects and larger animals. And they have no choice but to protect themselves against the most harmful forces of the climate, including sudden shifts in weather that occur with each passing day and more gradual changes as one season blends into the next.

When animals need food they go searching for it. Most

1

plants make their own food by the process of *photosynthesis* (pronounced foto-*sin*-the-sis), relying only on carbon dioxide in the atmosphere, available water and minerals in the surrounding soil, and light that is not always plentiful. For example, low-growing plants in dense forests may have to make do with brief bursts of energy captured from flickering spots of sunlight that penetrate the overhead leaf canopy.

Ways to survive the rigors of the environment are inherited by and built into each creature. Plants that grow best in a desert, on a mountain, in the sea, or tangled with others in a tropical jungle, for example, are those that are best adapted to such *habitats* (the specific places where plants and animals live).

Each habitat has its own environmental characteristics— high or low temperatures, bright light or shade, plentiful rain or drought, fertile or poor soils, to name only a few. And plants that have found a place in a habitat have inherited their own methods for surviving these conditions.

Such methods are custom-made for each type of plant. You would not expect trees that grow well in jungles to last more than a short time in a desert. And a desert cactus could not live in a jungle, on the snow-swept slopes of a high mountain, or under water in a pond.

Plants making their homes in such places are ones that, over millions of years, have become well adapted to the habitat's environmental conditions. And as long as those conditions remain unchanged, or change very slowly, the adapted plants survive. But when rapid environmental changes occur, such as those brought on by modern civilization, many plants die because they are unable to make an equally rapid adjustment.

Survival is the principal work of all living things. It is the driving force upon which life depends. For *species* (the different kinds of plants and animals) to continue, individual members must live long enough to reproduce. By creating offspring, a plant or animal adds to the unbroken stream of life that, over billions of years, has spread over the world we live in.

Each species possesses survival tactics that originated in ancestors millions of years ago—in the days of the dinosaurs, or even earlier. Using some of their survival methods, plants exploit the habitat's resources by sending out roots in search of water and nutrients, and spreading their branches to capture the sun's rays. Other inherited traits give plants the power to defend themselves against their habitat's most destructive features. Plants have uncanny ways of dealing with such threats as excessive cold or heat, water shortages, and animals looking for a meal.

At times environmental conditions seem to conspire to prevent plants from growing rather than help them enjoy long and healthy lives. But they do grow, occupying almost every part of the earth's land surface and the lakes, seas, and oceans that make our planet unique in the solar system.

This book looks at some survival secrets of plants in such diverse regions as the arctic, deserts, and tropical rain forests. Although the plants in these habitats struggle to survive, in even the most ordinary places the plants need effective methods for coping with other environmental challenges.

The science of *ecology* looks closely at the relationships between plants, animals, and the environment. Discoveries that ecologists have made over the years help us peek into the private lives of plants, see how they interact, and learn how each fits into the elaborate but fragile pattern of life on earth.

THE FOREST'S ENDLESS CYCLE

CHAPTER I

SURVIVING THE CHANGING SEASONS IN A DECIDUOUS FOREST

The Temperate Zones

The *temperate zones* lie to the north and south of the tropics, which girdle the earth at the equator. In the temperate zones, fall, winter, spring, and summer mark the passage of the yearly cycle—each season presenting different weather conditions that plants must survive. Winters are generally cold and wet, and frequently snowy. Summers become hotter and drier as the months go by. Spring and fall are periods of mild weather that the plants take advantage of to begin and end their growth.

In spring, as the last snows melt, the soil is saturated with water. Air temperatures rise as the sun climbs higher in the sky, and each passing day becomes longer. Temperate-zone plants that have slept through the cold winter months burst into springtime activity.

Fresh, new leaves pop from branches that stood bare all winter. Some trees, such as cherry and peach, cover themselves with spectacular displays of flowers, even before the first leaves emerge—better to be seen by insects that come to collect pollen from the flowers and so help the plants reproduce.

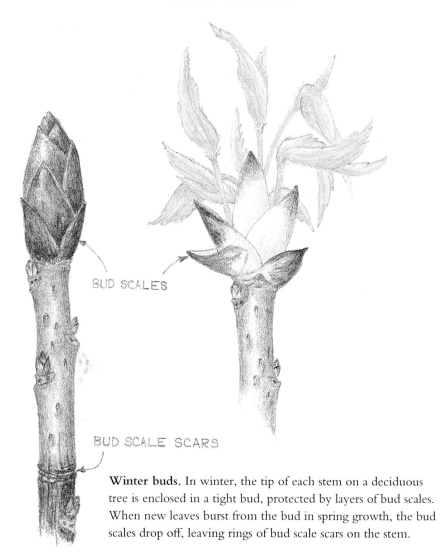

BUD SCALES

BUD SCALE SCARS

Winter buds. In winter, the tip of each stem on a deciduous tree is enclosed in a tight bud, protected by layers of bud scales. When new leaves burst from the bud in spring growth, the bud scales drop off, leaving rings of bud scale scars on the stem.

By autumn, the sun begins to sink closer to the horizon, nights are cooler, day lengths shorter. The soil's water supplies are becoming depleted and first frosts powder the landscape in early mornings—a warning that winter is on its way. It is time for animals to seek shelter and secure their food supplies. And it is time for many plants to enter their annual period of *dormancy*, similar to the hibernation of some animals.

When plants become dormant they stop growing, their liquid, food-laden sap stops flowing, and each branch's tender growing tip is carefully enclosed in a tight, frost-resistant bud. As a colorful prelude to the plants' entering winter sleep, leaves turn from green to yellow, orange, or red before dropping one-by-one to the ground. Later, you will see how these color changes occur.

What Is a Deciduous Forest?

Deciduous (de-*sid*-u-us) plants lose all their leaves each fall; examples of deciduous trees include maple, birch, beech, and elm, all of which are classed as *broad-leaved* species because

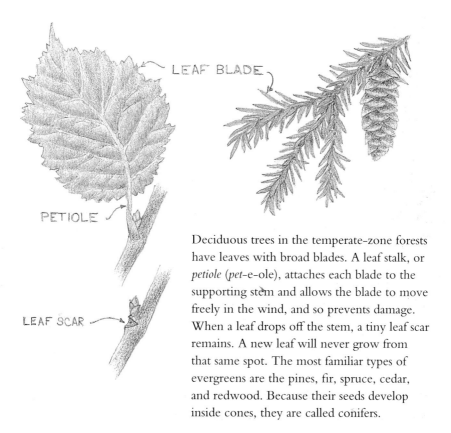

LEAF BLADE

PETIOLE

LEAF SCAR

Deciduous trees in the temperate-zone forests have leaves with broad blades. A leaf stalk, or *petiole* (*pet*-e-ole), attaches each blade to the supporting stem and allows the blade to move freely in the wind, and so prevents damage. When a leaf drops off the stem, a tiny leaf scar remains. A new leaf will never grow from that same spot. The most familiar types of evergreens are the pines, fir, spruce, cedar, and redwood. Because their seeds develop inside cones, they are called conifers.

7

their leaves have broad, flat blades. The leaves of *evergreens* remain on the branches throughout the year. Pine, fir, and cedar trees have narrow, needlelike leaves; other evergreens, such as holly, are broad-leaved.

In the northern temperate zones of Europe and North America, especially in the eastern parts of the United States and Canada, vast forests of broad-leaved, deciduous trees once covered the land. Exploited by people for fuel and everything from furniture to wooden ships, over the past centuries these forests have dwindled to only a shadow of their former glory. But enough small sections of deciduous forest remain for us to study the intricate seasonal growth cycles of various plants in these communities.

Preparations for Winter

Why do deciduous plants lose all their leaves each fall? Try placing a broad-leaved plant in a freezer for several days, and then bring it out to thaw. Although the stems may not show immediate damage, the leaves become limp and soggy, unable to recover their original shape and no longer useful for the plant.

During freezing, as water in plant cells turns to ice, the sharp crystals expand and puncture cell membranes, creating havoc in the once orderly structure of the leaf tissues. Frozen and thawed leaves on a tree quickly become sites where bacteria and fungi (molds and mushrooms) invade the stems.

How much better for plants to drop their leaves before freezing damage can occur, and to set a protective seal over the *leaf scars*. (These are left on the stems at places where the leaves were once attached.) The process of separation is called *leaf abscission* (ab-*sis*-yun).

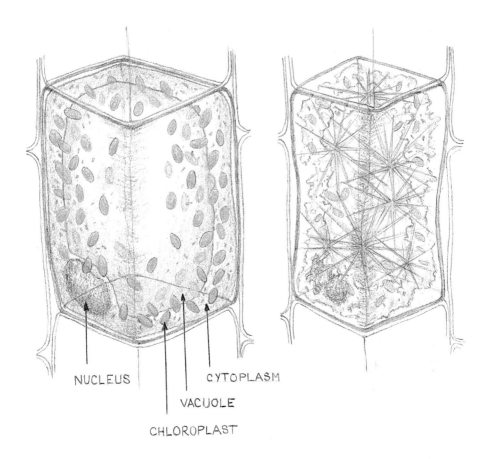

NUCLEUS   CYTOPLASM

VACUOLE

CHLOROPLAST

When water in a cell turns to ice, the cell's orderly structure is damaged beyond repair. In normal cells, water is contained in two sacs surrounded by delicate, transparent membranes. Numerous small bodies are suspended in the outer sac, or *cytoplasm* (*site*-o-plas-em). A nucleus controls the cell's many activities. In the cells of green stems and leaves, *chloroplasts* (*klor*-o-plasts; green bodies) contain the pigments and substances needed for photosynthesis. The cytoplasm is the living part of a cell. Inside the cytoplasm and filling a large part of the cell, a *vacuole* (*vac*-u-ole) contains most of the cell's water. In the living cell, the vacuole acts as a storage tank for water, food, and chemical wastes. When ice crystals puncture the vacuole's membrane, the waste substances flow into the cytoplasm and kill it.

How Leaves Lose Their Hold on the Plant

The timing of leaf abscission and entry into dormancy is regulated by chemical substances called *plant hormones*. Plants produce hormones in response to environmental cues provided by the changing seasons, such as the decreased temperatures and shorter days of autumn.

Inside and at the base of each leaf stalk, a layer of cells—the *abscission zone*—can be seen with a microscope. Within those cells a hormone causes cell walls and pectin to dissolve away.

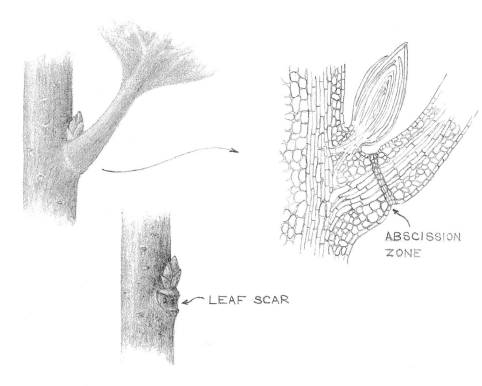

ABSCISSION ZONE

LEAF SCAR

In fall, leaves separate from stems at the base of their leaf stalks, or petioles. In a narrow band, called the abscission zone, the cells no longer stick together. Under a microscope (diagram, right), the abscission zone is easy to see. After the leaf has dropped, the surface of the leaf scar is sealed with a layer of cork cells.

10

Pectin is a "glue" that binds the walls together. As the joint between leaf and stem becomes weaker, the leaf is prepared to be blown away by the first autumn breeze.

The abscission process ends when the leaf scars are covered with a layer of protective *cork*. Cork blocks entry by harmful fungi, and prevents water loss from the stems.

The Forest's Nutrient Cycle

When leaves change color before abscission, a process critical for survival is happening inside. Minerals such as nitrogen and magnesium are drained out of leaf tissues and relocated in stems where they can be used the following spring for growing new leaves.

Minerals that are not transferred to stems are carried to the ground in the falling leaves. These minerals eventually return to the plants, but only after the leaves have rotted and roots have reabsorbed the minerals from the soil. The fall of leaves each year is an essential part of a deciduous forest's mineral nutrient cycle.

How Plants Avoid Frostbite

When leaf abscission is under way, plants are busy making other preparations for surviving winter. To stop water in stems from freezing, the plants fill their cells with "antifreeze" consisting of a concentrated sugar solution. When so fortified, only at abnormally low temperatures does water turn to ice. The antifreeze we use in car radiators in winter works the same way, but has a different composition.

Some of the plants' water may be stored in spaces between cells where, if it freezes, the rigid walls surrounding each cell

The mineral cycle. When fallen leaves rot on the ground, minerals are washed into the soil. Decaying branches release their stored nutrients more slowly. In spring and summer, roots draw the minerals back into the tree where they are used by the growing plant.

keep the ice crystals away from delicate cell membranes. Such preparations are part of the process of *cold hardening*.

With the arrival of spring, dormancy is broken when other plant hormones reverse the procedures described above. Water floods back into the cells, diluting the sugar solutions; the cells' chemistry (their *metabolism*) speeds up; and winter buds shed protective scales as pale green leaves unfurl into the sunlight. All such activity is set in motion by increases in air temperature and day length as winter slowly turns to spring. Even when dormant, the plants are remarkably sensitive to their surroundings.

The Forest's Layers

The leaves of plants in a forest form layers through which sunlight passes before reaching the ground. When all the leaves have grown, the largest trees intercept most of the light (95 to 99 percent) with wide-spreading *leaf canopies*. Because these trees shade the space below them, and thereby modify the environment on the forest floor, they are the *dominant* plant species in the forest. Dominant species also absorb the largest portion of available water and mineral nutrients from the soil. But for being the biggest plants in the forest, the dominant trees pay a price—they must take the full strength of the sun and suffer the greatest impact of heavy winds.

Sheltered below the dominant trees, the leaves of *shade-tolerant* bushes, ferns, and mosses occupy various levels in the forest structure, and survive summer with minimal amounts of light. The only direct sunlight they may receive is in short flashes, called sun flecks, when wind briefly parts the overhead leaf canopy.

13

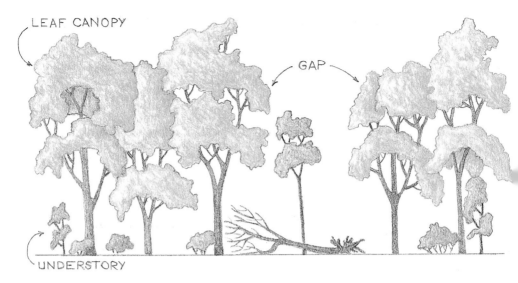

LEAF CANOPY

GAP

UNDERSTORY

In summer, below the forest's spreading leaf canopy, low-growing plants grow slowly in the dim light of the understory. When an old tree dies, a gap is opened in the canopy. Young trees grow rapidly in the gap's bright light and soon fill the space. Shade-tolerant plants, which do well in the understory, may be killed by bright sun that floods the gap. Such species will not return until the leaf canopy has grown and the gap is closed.

Growing among these plants, saplings of the dominant species make slow yearly progress until gaps, opened in the leaf canopy by the death of parent trees, allow full sunlight to reach the saplings' leaves. All of the plants growing below the dominant trees' leaf canopy make up the forest *understory*.

Most of the plants in the temperate-zone, deciduous forest are *perennials* (plants that, once established, grow and reproduce for many years). The forest supports few *annuals* (plant species that germinate from seeds, reach maturity, and reproduce in one growing season). In the forest understory, the growing season is cut short when the overhead leaf canopy blocks out the sun. For most annuals to complete their life cycles, the springtime period when full sunlight reaches the forest floor is all too short.

Fast Growth in the Understory

In winter and early spring in a deciduous forest, the trunks and bare branches of dominant trees absorb only a small portion of the sunlight, allowing as much as 70 percent, or almost three-fourths, of the sun's energy to reach the soil. In the short period between the end of winter and closure of the leaf canopy in early summer, the understory plants take advantage of the bright light, warm temperatures, and plentiful moisture and nutrients. For some, this may be their only period of growth; for the remainder of the year they stay dormant.

In only two or three months they must add to the size of their roots and stems, make new leaves, and even complete reproduction, which includes the complicated steps of forming flowers, fruit, and seeds, and of seed dispersal.

Shade-tolerant understory plants do not have to work so fast because their ability to photosynthesize in the shadow of the leaf canopy permits them to extend growth and reproduction into summer. But if sun-loving perennials in the understory are to complete their annual growth in the short time allotted to them before the canopy closes, they must get an early start. Even before the last snows melt, the plants break dormancy, all the while risking possible damage by late frosts.

The Forest's Feeding Plan

Energy to sustain early growth comes from foods that the plants produced and stored the previous spring. That is, before being cast into the canopy's deep shade, the understory must make and store new food supplies for use the following year. These plants are always preparing for their needs one year ahead.

Although food is made by photosynthesis in leaves, it is mostly stored underground in roots or other structures, including *rhizomes* (special underground stems) and *bulbs*. A bulb is a tight package of fleshy leaves attached to a short stem;

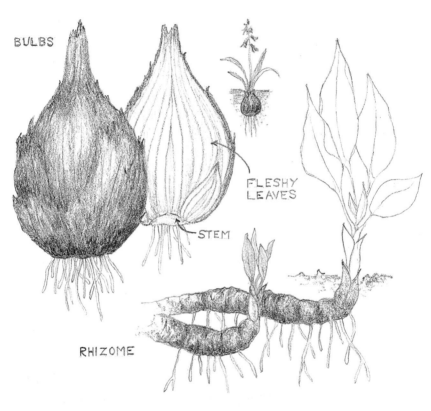

BULBS

FLESHY LEAVES

STEM

RHIZOME

Bulbs and rhizomes. These unusual stems grow underground where they store food and help the plants survive winter when their above-ground parts die back. After remaining dormant throughout the coldest months, new leaves and flowers emerge from protected growing tips. Rhizomes consist of stem tissue. Bulbs have layers of pale, fleshy leaves clustered around a short, flat stem. New bulbs develop from buds inside the parent bulb.

onions are familiar bulbs. Because most of the forest plants depend on stored food for springtime growth, they have to secure it throughout the year against hungry animals. What better way than to hide the food underground?

For their first stages of spring growth, the dominant trees depend as much as the understory plants do on stored foods. Newly formed leaves on the trees are poorly prepared for photosynthesis until they have expanded to almost their full size.

Even before their winter buds open, forest trees are pumping food in the form of a sugary liquid from roots to branch tips. At this time of year, the bark of sugar maple trees can be cut to drain some of the liquid sap out of the trunk. Boiling the sap thickens it into maple syrup.

A Chance for Seeds to Germinate

Fortunately for the understory, the dominant trees sleep until later into the spring. By the time the trees arouse themselves, the understory has enjoyed several weeks of early springtime growth, and seeds that survived winter in the soil have had an opportunity to germinate.

Seed germination requires warm soil temperatures and plentiful water, both of which occur at this time of year. If, by chance, seeds should germinate in summer, the dim light on the forest floor often causes *seedlings* (newly germinated, baby plants) to grow with spindly stems and poorly developed leaves. Their chances for survival are slight.

The best time of year for seeds to germinate in the forest is in early spring. By the time the canopy closes, the young plants are strong, well established, and able to survive for several months in the shade.

17

When a seedling grows in full sun, its stem is short and thick, the leaves are closely spaced, and they are green and fully expanded. The same plant in heavy shade has a thin, stretched-out stem bearing small, pale leaves. This is called an *etiolated* (ee-tee-o-late-ed) condition. Etiolated plants are weak and soon die.

Plants spend a lot of time and energy making seeds, because it is from them that the next generation grows. Many forest species have seeds with a built-in system that promotes germination in spring and prevents wasteful germination under the canopy in summer. This system depends on red light reaching seeds on the forest floor.

Sunlight is composed of the various colors seen in a rainbow. When light enters a leaf, the red and blue parts of the visible spectrum are trapped by leaf pigments. (The pigments channel the light's energy into food production in photosynthesis.) Under a heavy leaf canopy, very little red light reaches the ground, but green wavelengths pass right through the leaves. That is why leaves look green.

Many seeds possess special chemicals that, when stimulated by red light, start the seeds' germination. This happens in

springtime in the forest, before the dominant trees have
opened their new leaves. When the sun's red wavelengths
reach and penetrate into the soil, the waiting seeds are stirred
into activity.

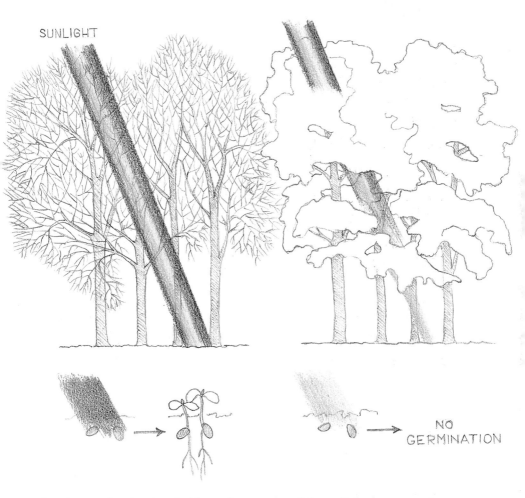

SUNLIGHT

NO
GERMINATION

In winter and spring in a deciduous forest, most of the sunlight (represented
here by a color spectrum) reaches the soil between the trees' bare branches.
When the snow melts and the soil warms in spring, red wavelengths in sunlight waken dormant seeds. In summer, the leaf canopy filters the sunlight.
Very little red light penetrates to the forest floor. The green wavelengths that
do reach the ground have no effect on seed germination.

The Trees' Summer Growth

If spring is the season of growth in the forest understory, summer belongs to the dominant trees. Their fully expanded leaves overlap each other to form a huge, green umbrella, held high above the forest floor by spreading branches. In such a position, the leaves capture most of the sun's energy that falls on the forest, quickly converting it into food by way of photosynthesis.

The dominant trees' large size demands proportionally large amounts of food to sustain the chemistry of living cells, to promote new growth, and to make flowers, fruit, and seeds before summer's end. A portion of the food produced in the leaf canopy in summer is shipped to the roots for storage, and made ready for breaking dormancy the next year.

Underground, roots begin growth in summer and may continue working their way into the soil during fall. Throughout summer, branches grow longer at their tips, and tree trunks thicken as layer upon layer of cells are added to the wood and bark tissues.

Each year, wood builds up inside the trunk in a distinct *annual ring*. Each ring records the tree's springtime and summer growth. Wood production stops in fall and winter. Counting the annual rings in a tree trunk gives us the plant's exact age. And the width of each ring tells us about rainfall patterns— narrow rings in dry years, wide rings in wet.

The Forest's Water Cycle

Water, drawn from the soil by tree roots, rapidly moves through cells in the outer part of the wood (the *sapwood*), up the trunk and branches, and into the leaf canopy. Then warmed by the sun, water in each leaf turns to vapor and passes

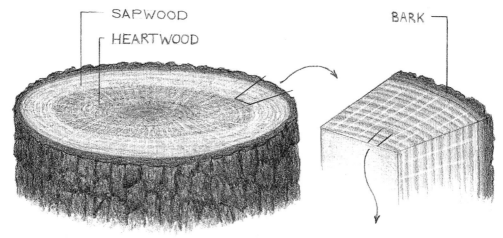

In a tree stump, you can see the difference between bark on the outside and wood forming the core of the tree. In some trees, the wood has two shades—light sapwood toward the outside, darker heartwood at the center. Water flows upward through the tree's sapwood. Both sapwood and heartwood are marked by annual rings that tell the age of the tree. Each annual ring consists of microscopic but relatively large-diameter cells that were produced during spring growth, and smaller cells formed in summer. There is no growth in fall or winter. The different cell sizes can be seen with a microscope, as shown in the lower diagram. New rings are always added to the outside of the wood. In the close-up view of portions of two rings, the inner ring (1) developed first; the outer ring (2) is the more recent.

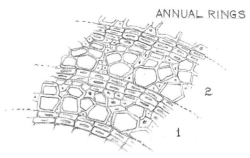

into the atmosphere through thousands of tiny pores, or *stomata*, in the leaf surface. The process of water vapor loss from leaves is called *transpiration*.

Almost all (about 98 percent) of the water taken up by roots is lost in transpiration. On a warm day, a large maple tree, for example, may lose 50 to 60 gallons (160 to 250 liters) of water into the atmosphere each hour. An acre of temperate-zone, broad-leaved trees has been estimated to transpire about 8000 gallons (30,000 liters) of water in one day. Even on a hot, dry, midsummer day, the air below a forest canopy feels cool and

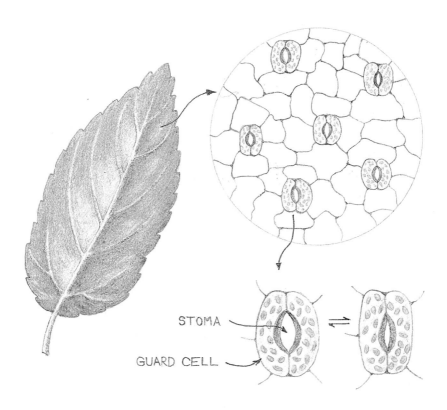

STOMA

GUARD CELL

Thousands of tiny holes, or stomata, dot the surface of a leaf—most often the lower leaf surface. Each stoma is surrounded by two guard cells. These can easily be seen under a microscope because they are the only cells in the leaf surface that contain chloroplasts. Guard cells control the opening and closing of the stomata. Opening occurs when the plant pumps water into the guard cells. As they swell, the thin, outer wall of each cell stretches more than the thick, inner wall. And so the two cells curve away from each other and open the stoma. When water is pumped out of the guard cells, they relax and close the stoma.

moist from the water vapor that the plants have passed into the atmosphere.

Plants cannot avoid transpiration when they open their stomata to take in carbon dioxide for photosynthesis. Although it may seem like a wasteful process, transpiration does have a useful and important side effect. When water vapor passes into the atmosphere, a suction force is established in the plants' water-conducting system—a force strong enough to pull water and minerals out of the roots and carry them to the uppermost leaves and branches.

With so much water being drawn from the soil by roots and lost from leaves, the soil becomes dryer and dryer until, by late summer, a shortage of water in the forest causes the dominant trees to slow their growth. In years of drought, when there is little rain to replenish depleted water supplies, the trees may begin early leaf abscission and prepare for dormancy even before summer has officially ended.

But if the forest canopy does persist until autumn, the cool night temperatures and shorter days of September and October are sure to start the season's normal color changes in leaves and to trigger their falling from the trees.

Pigments Unmasked

Leaves change color when the green pigment, chlorophyll, decomposes in leaf cells to reveal orange and yellow pigments, present all summer but hidden from sight by the more abundant chlorophyll.

In some trees, the unmasking of the yellow-orange pigments is accompanied by production of brilliant red ones, made from sugars and other substances in the leaves. The purpose of this last-minute display of added color is not known.

CHLOROPLAST   VACUOLE

How leaves change color. In a green leaf, both green chlorophyll and yellow-orange carotenoid pigments are contained in the tiny chloroplasts. Because there is more chlorophyll than carotenoid, the leaves appear green. In fall, after the chlorophyll decomposes, only carotenoids remain to give the leaf its glowing, golden color. In some plants, the leaf cells produce red pigments, *anthocyanins* (an-tho-*sy*-an-ins), that are stored in the vacuoles. As the anthocyanins collect and mask the carotenoids, the leaf turns red.

Winding Down at Year's End

Soon the dominant trees shed their spent leaves, allowing more and more direct sunlight to reach the ground. The understory plants are offered a brief period for active photosynthesis and renewed growth, before taking their turn to enter winter sleep.

By late fall, the climate begins to change for the worse as strong winds, heavy rainfall, low temperatures, cloudy days, and long nights usher in winter. Although the forest plants are facing the most destructive weather conditions of the year, they are well prepared for survival. Tree trunks and branches are insulated against frost by layers of cork in their outer bark; living cells are protected with sugary antifreeze; and tips of branches are secure in dormant winter buds.

Some low-growing plants on the forest floor let their leaves die back, leaving only bulbs or rhizomes to spend winter underground. Others find protection against wind and frost under a blanket of snow.

The forest deceives us when it appears so lifeless in winter. Although deep in sleep, the plants are still sensitive to changes in their environment, especially temperature changes. They are also measuring the length of days, recognizing differences between the short daylight hours of midwinter and increasing day lengths when winter turns into spring.

Another Year, Another Growth Cycle

To ensure their scheduled opening in early spring, dormant buds must experience several weeks of winter cold, at or below freezing, without which the buds fail to open. And for seeds from many of the forest plants to germinate in early spring, they too must first endure an extended period of low winter temperatures. Researchers are still trying to find out how chilling in winter promotes growth in spring.

By early spring, even before the last snow has melted, seeds may begin to germinate. And low-growing plants may start their growth unseen under the snow cover, using only the dim light that filters down to them. But some seem so eager to get a head start on spring that they push flowers through remnant snow banks to show off radiant colors in the sparkling sunlight.

The plants' springtime opening act, the green, shady forest in summer, and fall's colorful curtain-call may draw our attention away from the life-and-death struggle for survival going on behind the scenes. Year after year in the forest drama, the ability of plants to endure the changing seasons depends on:

❧ Taking full advantage of times when conditions are most favorable for growth (spring for the understory; summer for the dominant trees).

❧ Making sure the plants are completely dormant before winter arrives.

❧ Using seasonal changes in temperature and day length as cues to "turn on" leaf abscission, dormancy, or springtime growth.

❧ Storing plenty of food during periods of growth, ready for use early the following year.

Although the understory plants receive protection from the dominant trees against wind and the sun's heat, they must also live with changing conditions imposed by the larger trees' seasonal growth. As the leaf canopy expands, the environment

26

on the forest floor shifts from one filled with bright light to heavy shade, from being warm in spring to cool in summer.

Such conditions force the understory into slow summer-time growth, or even into early dormancy. Because of that, the understory plants can never match the yearly food production and growth gains achieved by the canopy trees. Held high on outstretched branches, the trees' leaves are well situated by summer to take advantage of the year's most intense sunlight for maximum photosynthesis.

Whether they are canopy or understory species, each is wonderfully adapted to its place in the forest structure. Each survives the changing seasons in its own special way. And each contributes to the intricate organization of a plant community called the deciduous forest.

FROM TREELINE, THE ARCTIC TUNDRA

ALPINE TUNDRA

CHAPTER 2

SURVIVING
THE WORLD'S WORST WINTERS

Arctic Tundra

If you were to travel northward in the temperate-zone deciduous forests, you would notice the broad-leaved trees gradually becoming mixed with evergreen *conifers* (cone-bearing trees with leathery, needlelike leaves). Eventually you would see only conifers such as pine, fir, spruce, and hemlock. Vast forests of these trees form an ecological community called the *taiga* (*ty*-ga, a Russian word) that stretches in a broad band across northern Europe, Russia and Siberia, Canada, and Alaska.

Traveling to the taiga's northern limits, you would see conifers that grow shorter and in less dense stands than those farther south. And as you approached the arctic region, even fewer trees would dot the landscape. At *tree line*, they disappear altogether. Here taiga merges into the bleak, open spaces of the *tundra*, stretching to the Arctic Ocean at the top of the world.

As unfriendly as the arctic tundra may appear at first, a close look at the land reveals a rich carpet woven from an

The delicate, leafy stems of mosses often grow in tight clusters or form spreading mats on the soil surface. Capsules, suspended on slender stalks, are ready to scatter their spores.

Two different types of plants live together in the body of a lichen. Under a microscope, green algae cells can be seen among a tangle of fungus threads. The algae make food by photosynthesis; the fungus supplies minerals and acts like a sponge to hold water. Some lichens grow flat on rocks; others grow upright and are intricately branched.

30

astonishing number of plant species. Chief among them are many types of grasses, mosses, and lichens. The tundra plants are especially interesting because their story of survival shows us how living creatures have adapted to one of the most extreme climates on earth.

Frozen Soils and Endless Days

A unique feature of the arctic tundra is the presence of *permafrost*—permanently frozen soil as close as 10 to 20 inches (25 to 50 cm) below the surface and sometimes extending hundreds of feet underground. Above the permafrost a relatively thin layer of soil is either frozen in winter or slowly thawing in summer.

Because the tundra plants' roots cannot grow into the permafrost, they are forced to occupy the upper few inches of the surface soil—the first to thaw in late spring. Only as ice in the soil turns to water can roots use it. In such shallow topsoil, trees and bushes, which normally have deep root systems, find it difficult to become established.

When winter winds sweep unchecked across the open tundra, they carry sharp, abrasive ice crystals. Blasted with the blowing ice, any plant daring to raise its branches above the thin snow pack is quickly cut down. You can understand why tundra plants are all low-growing, rarely exceeding 6 inches (15 cm) in height.

To prevent their being blown away, these plants are well anchored to the soil with spreading roots, which are often four times greater in volume than the above-ground shoots (stems, leaves, and flowers). Underground food storage struc-tures are common and can take different forms, including

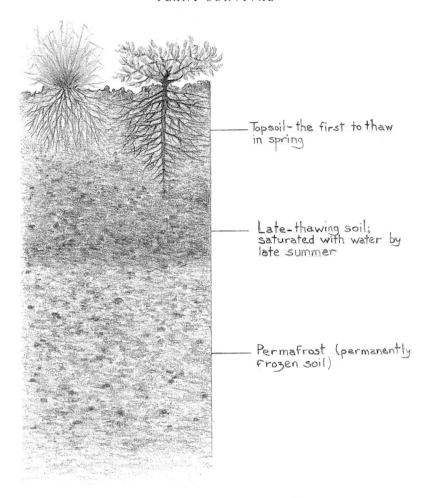

Topsoil - the first to thaw in spring

Late-thawing soil; saturated with water by late summer

Permafrost (permanently frozen soil)

The roots of arctic tundra plants grow best in the shallow layer of topsoil that warms in the springtime sun. As thawing continues deeper into the soil in summer, the underlying permafrost prevents the water from draining away.

rhizomes and *bulbs*, thickened roots (like carrots), and fat *tubers* (like potatoes, although potato and carrot are not tundra plants).

The tundra offers its plants only a short season for growth— 50 to 90 days in June, July, and August, after the topsoil has thawed and air temperatures have climbed above freezing. Year by year, the actual length of the growing season varies, depending on prevailing weather conditions. With such an unreliable schedule, the plants have no time to waste after breaking dormancy: starting growth, making food, and even preparing for another growing season, twelve months away.

One advantage to life in these far northern regions is that plants enjoy midsummer days lasting 24 hours; here the "midnight sun" never sets below the horizon. In contrast, for several weeks in midwinter the sun never appears, but by then the plants are fast asleep in dormancy. Although the tundra's summer season for plant growth may be relatively short, abundant sunlight provides all of the plants' annual energy needs.

Trapping Heat in Leaves and Flowers

Most tundra plants are perennials. The few species of annuals, which must complete their life cycles from seed germination to seed production in two or three months, are all very small in size.

Like rounded, green pillows sitting on the soil surface, the above-ground shoots of long-lived perennials often form *tussocks* or *cushions*—tight clumps of stems and leaves. Such forms help to protect the plants against damage by cold and wind. Each low-growing plant cluster absorbs and holds so much of the sun's heat that leaf temperatures have been recorded as much as 36°F (20°C) higher than the surrounding air. Elevated leaf temperatures promote photosynthesis.

Just as ingenious are the flowers of some tundra plants. Their cup-shaped, highly reflective petals concentrate sunlight onto the center of each blossom, raising its temperature 4 to 18°F (2 to 10°C) above the temperature of the outside air.

33

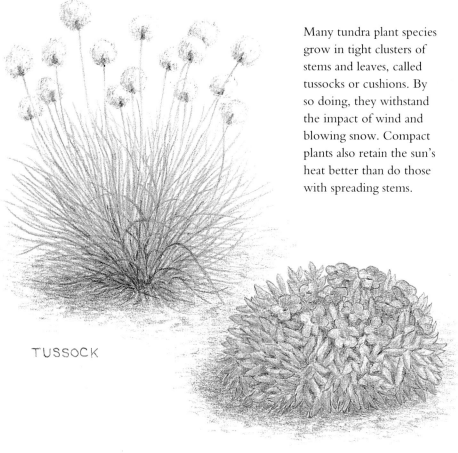

Many tundra plant species grow in tight clusters of stems and leaves, called tussocks or cushions. By so doing, they withstand the impact of wind and blowing snow. Compact plants also retain the sun's heat better than do those with spreading stems.

TUSSOCK

CUSHION

Insects, attracted to these warm pockets, collect the pollen and so help the plants reproduce. The added warmth also speeds up the development of seeds in each flower's ovary.

To maintain such a temperature difference between the inside of the flowers and the outside air, each day the flowers rotate on slender stalks. Preceding human invention by many thousands of years, the tiny blossoms act like dish-shaped antennas tracking the sun as it moves across the sky.

34

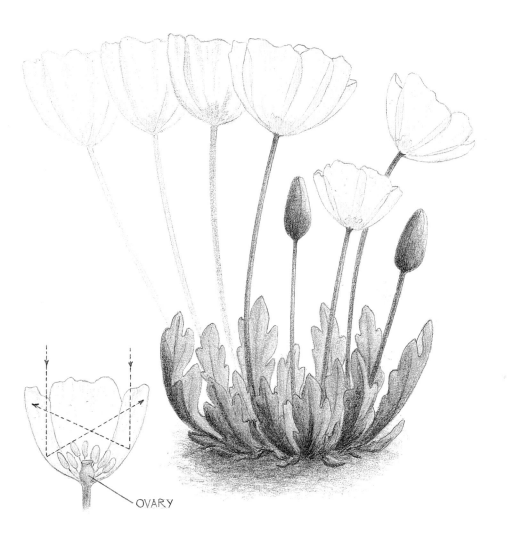

OVARY

Sun-tracking flowers. When the sun's rays reflect from the curved petals of cup-shaped arctic poppy flowers, the reproductive parts are warmed—including the ovary where seeds develop. Insects visiting the warm floral pockets carry pollen from flower to flower. To keep an even temperature throughout the day, the flowers follow the sun on rotating stems. The changing positions of one of the flowers are shown here.

35

A Harsh Climate and Soggy Soils

Despite the long days of summer, air temperatures in the arctic seldom exceed 60°F (16°C) on the warmest days. Compare this with the temperatures in your own part of the world. In most places, a 60°F day is considered fairly cool and hardly summer-like.

Throughout fall, winter, and most of spring, temperatures in the tundra stay below freezing, frequently dropping to less than −58°F (−50°C) in midwinter. The tundra's total annual precipitation (rain and snow) amounts to only 12 to 20 inches (30 to 50 cm), about half of which falls as winter snow. With so little snowfall, the snow is not as deep on the ground as you may imagine for such a cold climate.

A combination of low temperatures, sparse precipitation, and poor soil conditions challenges each and every tundra plant in its struggle to survive. Because the topsoil uses most of the heat it absorbs in summer to thaw the accumulated ice, roots must grow in soil temperatures that barely rise above freezing. And water from the melting ice can drain no deeper than the underlying layer of permafrost. The word *tundra* means marshy plain and very well describes this habitat's open expanses of soggy soils.

Even with so much moisture in the soil, dead plant parts rot slowly. It is different in the deciduous forest where minerals are cycled from plants to soil and back again. When fallen leaves and branches decompose on the ground the minerals are reabsorbed by roots. The rotting process, carried out by fungi and bacteria that populate soils, is most rapid in warm temperatures.

The opposite is true for the tundra's cold soils where slow

rotting accounts for a shortage of nutrients. Tundra soils are especially poor in nitrogen and phosphorus. In addition to having a meager mineral supply, the dense, marshy soils let in oxygen only slowly. As a result, they soon turn acidic. All such conditions combine to make growth difficult for most tundra species.

A Short Period for Growth and Food Production

If you visited the tundra in early spring, you would see few signs of plant activity. Only as roots begin to absorb water from the melting topsoil do dormant shoots slowly stir. Then in the lengthening days of late spring, three to four weeks after the plants waken from dormancy, flowers finally open from buds formed the previous summer.

Springtime flowering and the modest growth of new shoots use up most of last year's stored food. So in summer, the plants must take advantage of continuous daylight to meet continued nutritional needs and to restock underground food supplies for the following year.

With annual food production rarely exceeding the amount necessary to supply basic needs, the growth rates of tundra species are among the slowest in the plant kingdom. Even so, these plants alone possess the secret of survival in such a cruel climate: Only they can perform the countless chemical processes of living cells, including photosynthesis, at unusually low temperatures. Exactly how this feat is accomplished remains a mystery.

The short summer months impose an ambitious agenda on the plants. In addition to stocking up on food, they must also extend roots into the moist soil, absorb and store available

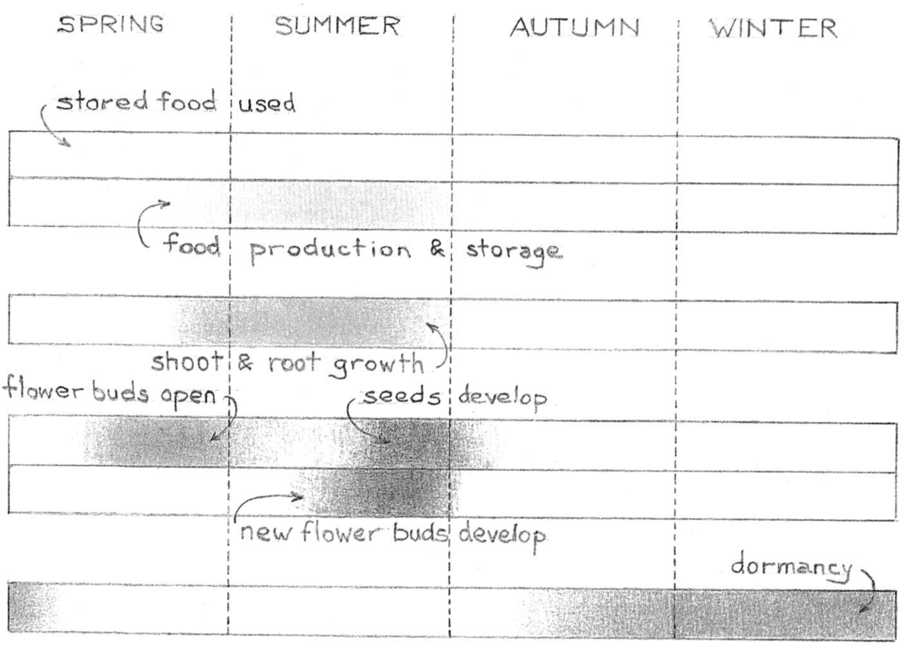

SPRING	SUMMER	AUTUMN	WINTER

stored food used

food production & storage

shoot & root growth
flower buds open
seeds develop

new flower buds develop

dormancy

With only a short spring and summer growing season, arctic plants work fast to complete their many tasks. Springtime growth, especially flowering, uses most of the food stored from the previous summer. Summertime food production also supplies energy for root and shoot growth, seed development, and the formation of next year's flower buds.

minerals, start work on next year's flower buds, and put the finishing touches to the current year's seed crop.

Fortunately, the seeds of tundra plants remain viable (ready to germinate) for many years, waiting for an occasional extra-warm spring to start to grow. *Vegetative reproduction*, by the branching of underground rhizomes or by forming new bulbs, for example, is the most common method of reproduction among the tundra's occupants.

Autumn comes to the far north before the end of August. A month later, first snows cover the land. The plants' entry into dormancy is prompted by a combination of lower temperatures, drying of the soils, and shorter days in fall as the sun sinks toward the horizon. Finally submerged in deep sleep, all of their life processes suspended, and their cells protected with sugary antifreeze, the tundra plants are well prepared to survive another long, harsh winter.

Alpine Tundra

Imagine being transported to the slopes of a tall mountain. Climbing upward, you would probably pass through dense forests before noticing that the trees are smaller and fewer in number as you approached *timberline* (tree line). Continuing your climb under the open sky, you would enter a habitat that closely resembles the arctic tundra. This *alpine tundra* is also populated by plant species whose biggest challenge is to survive most of the year in very cold temperatures.

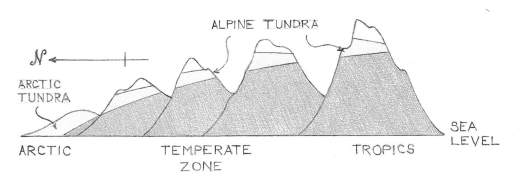

Tundra, shown here in yellow-green, extends from sea level in the arctic to alpine tundra at increasingly higher altitudes on mountain peaks, depending on how close the mountains are to the equator.

How high on a mountain must you climb to reach the alpine tundra? That depends on how far north or south of the equator the mountain is located. Only the highest peaks in the tropics have an alpine tundra near their summits. For example, in Africa you would have to be far up the slopes of Mount Kilimanjaro (19,340 feet, or 5895 meters, high) to find such a habitat. In northern Alaska, on the other hand, timberline and the presence of alpine vegetation may occur only a few hundred feet above sea level.

The principal difference between arctic and alpine tundras is the absence of permafrost in the mountain soils. The ground may become frozen in winter but thaws deep underground by summer, allowing most alpine soils to become well drained. Consequently, roots can grow to greater depths than in the frozen arctic soils. Such roots certainly are needed to anchor the plants on exposed mountain slopes where fierce winds tear at their branches.

Advantages to Low Growth
Small trees and bushes in such places grow stooped, their permanently bent stems and compactly trimmed branches all facing away from the direction of the prevailing wind. But most alpine plants avoid damage by wind and the crushing weight of the winter snow pack by simply growing close to the ground, often in the cushion forms so common in the arctic tundra.

Another advantage of low growth is that air temperatures are higher in a shallow layer 4 to 6 inches (10 to 15 cm) above the soil. The placement of stems, leaves, and flowers in the warmest air speeds up growth, photosynthesis, and reproduction.

40

On exposed mountain slopes, trees are shaped by the direction of the prevailing wind. Branches that try to grow on a tree's exposed side soon break off.

A Brief Opportunity for Growth

As in the arctic tundra, the season of plant activity high on a mountain may be limited to three to four months between snow melt and the early return of winter. Because summer night temperatures on mountains are colder than in the arctic, and daylight hours are shorter, alpine plants generally grow even more slowly than arctic species.

Compared with the arctic, mountain habitats are populated by larger numbers of annuals. But a majority of alpine species are perennials. Living in places where the growing season is so short, perennials have an advantage: After emerging from

41

dormancy, they simply add on to already well-developed roots and shoots, rather than growing complete, new plants from seeds each year, as do annuals.

Dangers in the Mountains' Thin Air

Plants living in the thin air of high alpine areas—several thousand feet above sea level—are bombarded with ultraviolet (UV) light. At lower elevations, the thickness of the earth's atmosphere filters out many of these harmful rays. Ultraviolet light can kill by destroying the delicate structures of living cells.

An effective method that some alpine plants use to protect themselves against the intense sun is to cover their leaves and young stems with dense mats of white hairs, giving them a silvery appearance and reflecting much of the light that strikes them.

The leaves of other plants contain *anthocyanin* (an-tho-*sy*-an-in), a red pigment. Anthocyanin absorbs UV rays before they have a chance to destroy chlorophyll and the photosynthetic system. Leaves containing a combination of red and green pigments are colored a dark purplish brown, almost black. Because dark objects rapidly absorb the sun's heat, the deep pigmentation imparts an added benefit—the leaves stay warmer than the surrounding air.

How Alpine Plants Manage Their Food Supplies

All living things must be supplied with food to stay alive. You eat several meals a day to give you energy for your activities. Plants don't eat; they make their own food by photosynthesis during daylight hours. They use some of the their food for immediate needs such as growth and the many complex processes that help the plants function. They store the

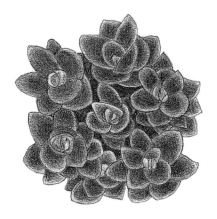

PROTECTION AGAINST UV LIGHT

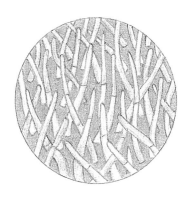

In the thin air on a high mountain, more of the sun's ultraviolet (uv) rays reach the ground than at sea level. To protect themselves against the harmful rays, the leaves of some alpine species contain a red pigment, anthocyanin. This acts as a uv filter. Anthocyanin combined with green chlorophyll gives the leaves a brown color. Other plants have leaves covered with dense mats of hairs that reflect light. Under a microscope, the hairs can be seen projecting from the leaf surface.

remainder for future use, especially for times when environmental conditions are unfavorable for photosynthesis.

In the arctic and alpine tundras, food storage is essential for survival. When the plants are dormant, they use very little food. But as soon as they awaken in spring, tundra plants require a lot of energy to get a jump-start on growth.

Springtime growth depends primarily on food made the previous summer and stored in roots, rhizomes, and other underground parts. Photosynthesis in summer replenishes those supplies. The amount of food prepared for the following year depends on weather conditions that vary from one summer to the next.

For example, clouds that frequently gather around mountaintops indirectly influence plant growth. In those summers when a heavy cloud cover hangs over the alpine regions, photosynthesis and food storage are reduced, compared to food production when the sky is clear.

Snow depth and melting rates also affect the growth of alpine plants. In years when unusually thick snow covers the ground, or a late snow melt delays the end of dormancy, too little time may be left in the growing season to completely rebuild the plants' stored food supplies. So the plants are forced to grow less the following year.

Strong winds keep the alpine plants trimmed close to the ground, dry out the soil, and blow away fallen leaves. If left to rot, the leaves could add precious nutrients back to the earth. But in exposed sites where the wind whips them away, a shortage of decaying vegetation results in soils that are poor in nutrients, and less favorable for plant growth.

In sheltered areas, on the other hand, the soils remain moist throughout summer, which promotes rotting of dead plant parts and the release of stored minerals. By midsummer, these favored mountain sites may be crowded with thousands of brightly colored wildflowers—a sharp contrast with what

After the snow banks have melted, the warm summer sun brings out the true colors of an alpine meadow.

seemed like lifeless meadows, slowly revealed by the melting snows only a few weeks before.

Compass Direction and Temperature Differences

Snow melts at different rates on a mountain's various slopes, depending on how much heat each receives as the sun makes its daily journey across the sky. Just as the shaded, north side of a building is cooler than the south side, on a much grander scale the north-facing slopes of a mountain remain the coolest throughout the year.

Patches of ground in the mountain's own shadow are the last to be uncovered as the snow melts in summer. And in places where the snow pack is deep, only the surface thaws before fresh layers of snow are deposited by the next winter storms.

During spring and summer, south-facing slopes are warmed in daytime to above-freezing temperatures. As the snow banks melt, first on ridges and later in sheltered gullies, vast quantities of water are released. Some sinks into the ground but most runs downhill—first in tiny trickles, then in streams that feed rivers making their way to the oceans.

When alpine plants are first exposed to the sun, after months under a heavy blanket of snow, many are flattened to the ground and hardly seem capable of recovering their normal shapes, least of all resuming growth and reproduction. But within a few days their leaves and branches spring back to life, and flower buds, formed at the end of the previous summer, quickly develop into full-fledged blossoms.

Slight differences in temperature occur between mountain-side *microenvironments* (small areas within the habitat). Even in the shelter of small rocks, plants bask in extra-warm pockets of

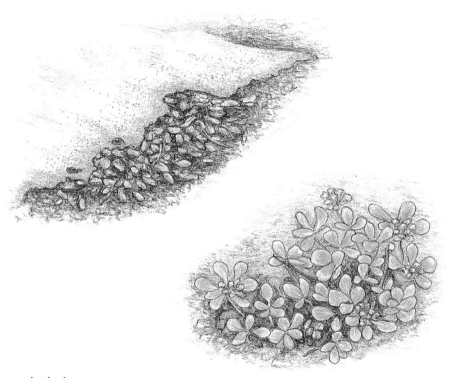

As the last snow melts in spring, the weight of the winter snow pack is clearly shown by the flattened alpine plants. But within a few days they spring back to life and make an early start to flowering.

air. And in shallow hollows in the ground, low-growing plants take advantage of warmer temperatures to stimulate growth in the summer sun.

Like living compasses, cushion-form plants first open their blossoms on their warmer, south sides, while on the north side of each cushion the flowers are still in bud. Likewise, as reproduction progresses, slightly higher temperatures on the plants' south sides favor a more rapid development of fruits and seeds.

47

Even a slight temperature difference between the south-facing and north-facing sides of a compact alpine plant is enough to stagger the opening times for its flowers.

Advance Preparations—The Key to Survival

Plants in other habitats normally finish one year's seed production before working on new flowers for the next year. Not so in the tundra. With such a short growing season each year, the plants cannot afford to waste a single moment of their waking days. So in the long days of summer, while many alpine perennials prepare to shed their latest crop of seeds, they begin to form flower buds, ready for blooming the following year. Such early preparations serve a useful purpose: In spring, after being released from the grip of the melting snow pack, the plants do not have to start the reproductive process from scratch.

Although most alpine plant species produce more seeds than arctic species do, the seeds may rest for many years without germinating. They wait for an ideal but unusual

48

spring season when the snow melts early and daytime soil temperatures reach about 70 to 85°F (20 to 30°C). After seed germination, most alpine perennial plants have to grow two or more years before being able to form their first flowers.

Shared Problems in Separate Regions

Plants of the alpine and arctic tundras have much in common. They spend long periods each year in a state of dormancy—the only way to survive low temperatures, the shortage of sunlight in the arctic winter, or burial under deep layers of snow on a mountain.

When snow and soil eventually thaw, the plants have only a short season for growth, stocking up on foods, flowering and making seeds, and getting a new set of flower buds started for the following year. Only when all of these activities have been completed can the plants safely go back into dormancy and survive another seasonal cycle.

Alpine and arctic tundra plants have special qualities that enable only them to exist in such harsh climates:

※ They metabolize, grow, and reproduce in the cool air temperatures of the growing season.

※ They have methods to prevent living cells from freezing in winter.

※ They withstand strong winds, the weight of accumulated snow on their branches, and, on high mountains, the damaging effects of ultraviolet light.

※ They quickly resume growth as soon as the snow cover and topsoil have thawed and daytime temperatures have risen above freezing.

🌿 They make enough food each summer to satisfy both their immediate needs and to store, ready for starting fast growth in spring of the following year.

Only a select group of animals and plants have evolved to live in the far north and on mountain peaks. Arctic and alpine tundra plants are so well adapted to survive the long, cold winters that when they are transported to more moderate climates, they grow poorly. While most of us would choose to avoid the tundra's extreme environment, native plant and animal species are perfectly at home in such a place.

IN THE RAIN FOREST

CHAPTER 3

SURVIVING COMPETITION IN A TROPICAL RAIN FOREST

Problems of Jungle Living

Tropical rain forests seem to be ideal places where, you may think, only the most fortunate plants are born. Jungle plants grow well with plentiful sunshine and all the water they could ever need. Temperatures never drop below freezing, and day lengths hardly change from one month to the next. In the tropics, years do not have a spring, summer, fall, and winter cycle. So plants are not restricted to certain seasons when they can grow, or by climate changes that force them into dormancy.

But jungle living is not as easy as most people imagine. The problem is that in such a perfect climate for growth, too many plants try to occupy the same space. As a result, they end up competing for available resources.

The countless thousands of plants in a tropical rain forest struggle to gain access to sunlight, to find scarce body-building minerals, and to claim enough room to grow. Only species with the best possible tricks for survival are winners in the crowded jungle realm. Yet, despite the strong competition, an unexpected harmony exists among many of the forest's inhabitants. Lofty trees let small intruders grow on their branches, and lowly plant forms provide life-giving assistance to mighty neighbors.

Jungles exist near the equator and where annual rainfall exceeds 80 to 90 inches (200 to 225 cm). In places where rain is restricted to certain "wet" seasons, alternating with "dry" seasons, the forest may be semideciduous—the leaves falling in response to drought rather than in response to short days and cool temperatures as in northern regions.

Year-round Growth

In the true rain forests of Central and South America, Africa, and Southeast Asia, plants grow throughout the year. Tropical trees produce wood continuously rather than in the distinct annual rings of temperate-zone species. Consequently, it is more difficult to determine their age. Leaf abscission simply happens when individual leaves get too old to function; and flowers are formed in periodic flushes, triggered by unknown stimuli, not in certain seasons of the year.

In the high temperatures of the tropics, leaves manage to stay cool by transpiration. With so much water vapor lost into the atmosphere from so many plants, plus frequent rain showers, the humidity in a jungle remains high, rarely falling below a steamy 95 percent in the forest's understory. There the air lies undisturbed except by the strongest winds.

Uncounted Numbers of Plants and Animals

Rain forests are home to the largest number and greatest diversity of plant and animal species of any place on earth. In a single jungle acre it is not unusual to find more than 100 different types of trees, compared with five or fewer tree species in a northern deciduous forest. And living on and around the trees, uncounted thousands of other kinds of plants add to the rain forest's lush vegetation. In the Amazon basin alone, one-third of the world's flowering plants (about 80,000 species) exist, along with thousands of species of ferns, mosses, lichens, algae, and fungi.

Scientists continue to discover animal species (especially insects) as they try to compile records of the rain-forest inhabitants and the roles each plant and animal plays in the complex workings of the jungle ecosystem.

The Rain Forest's Structure

Trees form the framework of the rain forest's intricate structure. Either high in their tangled branches or at various levels in the deepening shadow of the canopy, the forest trees create a host of microenvironments in which other plants can grow, depending on their individual preferences.

If you looked down from above, the forest canopy would be a sea of leaves. The canopy is supported by trees with flattened crowns of almost horizontal branches. Their sturdy trunks rise from the forest floor without branching for the first 60 to 80 feet (18 to 25 meters). And although the topmost leaves in the canopy may hang 100 feet above the ground, they are still overshadowed by the leaves of giant trees. Here and there, these huge trees poke through the canopy with enormous trunks and thickened limbs.

Many forest giants grow 200 feet (60 meters) tall, yet, like all jungle trees, are anchored by roots growing only a few feet

FOREST GIANTS

LEAF CANOPY

UNDERSTORY

A rain forest is composed of hundreds of tree species. Each spreads its branches and leaves at different heights above the ground, depending on the amount of light needed to survive. Forest giants and the tallest canopy trees absorb most of the full sunlight falling on the forest. Trees and bushes in the shady understory are adapted to the dim light.

into the soil. But the roots make up for this shallow foundation by spreading in dense mats across the forest floor. Some species of giant trees gain added support from *buttresses* (large, vertical flaps) that project from the base of their trunks. The

Even the largest forest trees are anchored to the soil with shallow, spreading roots that are interwoven with the roots of their nearest neighbors.

buttresses spread a tree's weight over a wide area and help to keep it from toppling when winds start to blow. A few plants support themselves with groups of *prop roots* around their bases, like poles straining to hold the stems in place.

The leaves of canopy and giant trees capture most of the sunlight that falls on a rain forest. These leaves are also exposed to the highest daytime temperatures, the greatest force of the wind, and lowest humidities in the jungle's many microclimates. To control evaporation of water, the leaves are smaller than on plants in the understory. They have a thick, outer waxy layer, giving them a smooth, glossy appearance. Rain that quickly runs off the leaves drips onto the underlying vegetation or is channeled down branches and trunks to the soil.

BUTTRESS ROOTS

Some species of rain-forest trees
keep themselves upright in the soft,
soggy soil by having unusual
supportive roots. Buttress roots
project from the tree trunks and
extend ribbonlike into the area
around the base of the tree. Prop
roots grow out of the stem and
angle down to the soil

PROP ROOTS

Plants Growing on Plants

The branches and trunks of canopy and giant trees, positioned as they are in the brightest light, support many *epiphytes*, which are plants growing upon other plants. The word *epi* means upon, and *phyton* means plant. Most epiphytes are sun-loving species, but too small to compete with forest trees by trying to grow toward the sun from ground level. Instead, they simply germinate from seeds or spores deposited on tree

Many species of mosses, lichens, ferns, and flowering plants grow as epiphytes on the trunks and branches of rain-forest trees. There they bask in brighter sunlight than is available on the forest floor. Of all the epiphytes, tropical orchids bear the most spectacular flowers.

branches by wind or animals. Epiphytes spend their lives in a sun-drenched environment but must share their supporting tree's fate when it eventually falls over and dies.

Roots of epiphytes cling tightly to the tree bark but do not penetrate it to steal food or water—they are not parasites. In fact, as mosses, lichens, and decaying leaf litter collect in a damp, thick mat among the larger epiphytes, some trees send out special roots from their branches to draw moisture and minerals from the mats they support.

While epiphytes share the treetops' abundant sunlight, they must also share the scorching temperatures and drying winds in such lofty perches. Because epiphyte roots never reach the ground, the plants' biggest challenge is to collect and store water and minerals for their daily needs.

Of all the types of plants that grow as epiphytes, orchids are the most spectacular with their brilliantly colored flowers of every imaginable shape. Orchids cling to their supports with special roots, some of which hang in clusters below the plants to soak up rain water in spongy tissues. Thick, fleshy leaves and stems store the collected water. Minerals absorbed by roots are washed out of dust and rotting leaves and other materials that collect around the plants.

In the jungles of South America, another group of epiphytes, the *bromeliads* (bro-*mel*-ee-ads), are common. Bromeliads are members of the pineapple family. One type, the "tank bromeliads," earn their name from the way the stiff, leathery leaves cluster tightly around each plant's short stem, forming a deep cup (or tank) where water collects. The leaves curve inward to direct rain water into the tank. On their

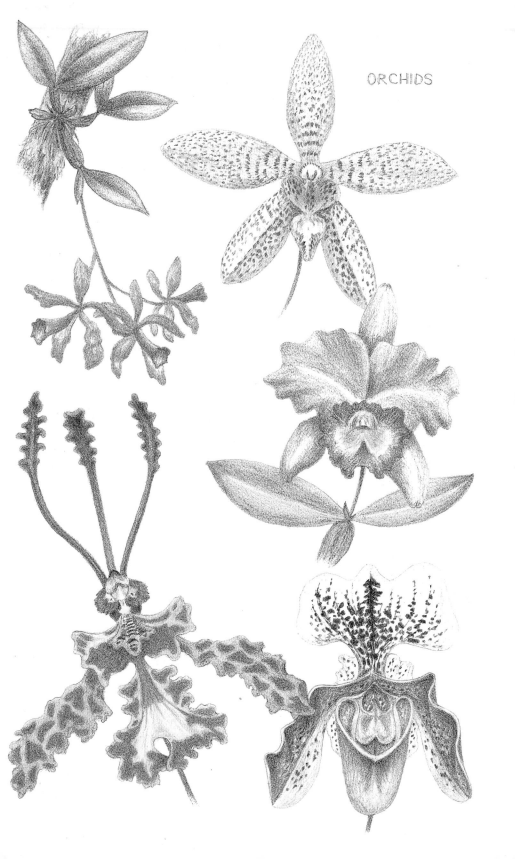

ORCHIDS

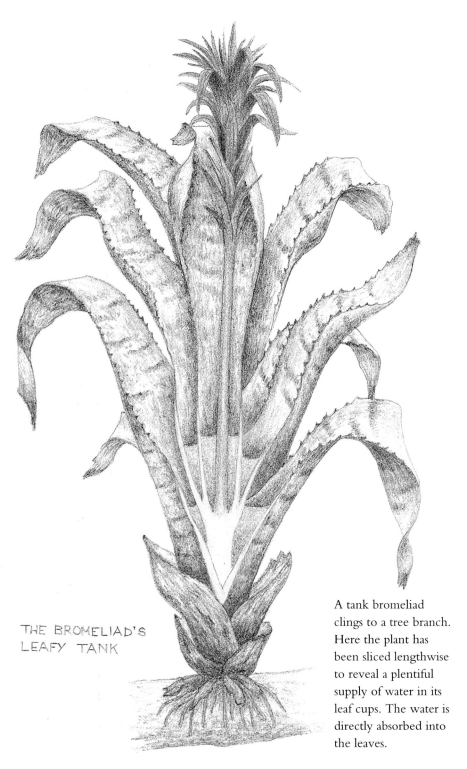

THE BROMELIAD'S
LEAFY TANK

A tank bromeliad clings to a tree branch. Here the plant has been sliced lengthwise to reveal a plentiful supply of water in its leaf cups. The water is directly absorbed into the leaves.

62

upper surfaces, special cells draw moisture and minerals out of the reservoir for the plant's use. Minerals are released from decayed leaves and other vegetable matter, the dead bodies of small animals, and animal droppings that all collect in the bromeliad tanks.

Tank size varies with species, some of the largest plants holding as much as 5.5 gallons (20 liters) of water. In these tree-top swimming pools, mosquitoes, frogs, and other small amphibians raise their young.

In the Grip of a Strangler Fig

Although the strangler fig starts life as an epiphyte, it soon reveals a more sinister side to its behavior. The fig's sticky seeds, having been carried into the forest canopy by birds or other animals, quickly germinate and send out several roots. Some cling to the tree branches while others rapidly grow down the trunks to the ground.

As more roots develop, they surround the trunk and fuse into a rigid cylinder that restricts further growth of the tree. Meanwhile, the fig's branches and leaves overshadow the tree's foliage, depriving it of light. Contrary to the fig's "strangler" name, the supporting tree most likely dies in the battle for light, soil nutrients, and water, rather than from being squeezed to death. Eventually, the dead tree rots away, leaving the fig free-standing.

Plants That Climb

The strangler fig grows downward to the soil, but numerous climbing vines grow from the ground, upward through the rain forest's dimly lit understory. Firmly grasping any plants

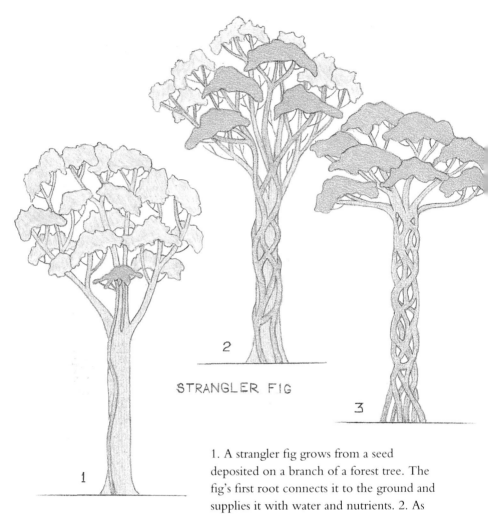

STRANGLER FIG

1. A strangler fig grows from a seed deposited on a branch of a forest tree. The fig's first root connects it to the ground and supplies it with water and nutrients. 2. As other roots grow and surround the tree trunk, the fig's branches and leaves overshadow the tree's foliage. 3. After the tree dies and has rotted away, only the fig remains.

strong enough to support their weight, the climbers struggle to reach the sunlight that floods the jungle canopy.

To locate a tree trunk suitable for climbing, the seedlings of some vines have an unusual but very effective strategy. Most young plants grow toward a light source for photosynthesis, but a vine seedling grows away from light, toward the darkest object in its surroundings—usually the shaded base of a large tree. Once it has reached the tree, the vine's stem changes direction and starts its climb up the side of the tree. The vine

The seedlings of most plants are strongly attracted to light. In the deep shadow on the rain-forest floor, the seedlings of climbing vines do just the opposite. They grow toward the closest dark object, most often the base of a tree. The vine stem then grows up the tree toward the sun.

saves energy by waiting to form most of its leaves only when the stem has emerged into the light.

The stems of *lianas* (lee-*anas*) hang from the upper branches of the forest giants and grow in great loops between canopy trees. Although the liana stems thicken and become woody, they remain flexible enough to sway in the wind, and are

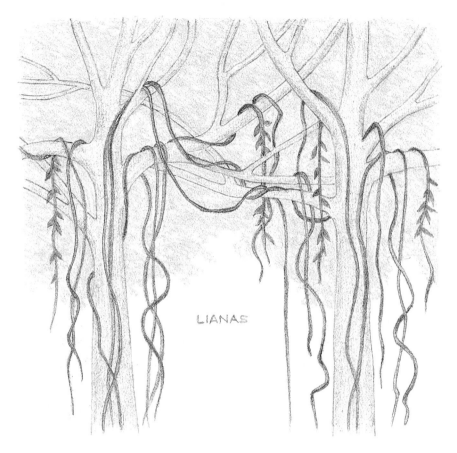

Lianas—woody, climbing vines—grow from tree to tree in twisted loops and hang from outstretched branches.

strong enough to form handy treetop walkways for the jungle's animal residents.

Lianas bind the tops of the forest trees together. Many times when a giant crashes to the ground, it and the connecting vines pull other trees with them. When this happens, the structure of the surrounding understory is destroyed and will not return to its original form for many years.

Life in the Understory

Understory plants exist in the year-round shadow of the forest canopy, in light intensities only a fraction (less than 5 percent) of full sunlight. So little light reaches the ground that few low-growing species occupy the jungle floor. Except for tree trunks and liana stems, the forest interior is surprisingly open and fairly easy to walk through.

The idea that jungles have dense, hard-to-penetrate vegetation comes from scenes along river banks and road cuts where full sun, reaching both canopy and understory, promotes the growth of tangled masses of vines, bushes, trees, and their epiphytes. But beyond such towering walls of stems and leaves, the forest floor is almost bare of plants.

Understory trees include shade-tolerant species that reach maturity only in the absence of full sun, slow-growing saplings of canopy and forest giant species, and dwarf palms. Trees in the understory have narrow, upward-reaching crowns compared with the spreading tops of the mature forest canopy.

A few species of low-growing *shrubs* (plants with several woody stems branching from ground level) and *herbs* (plants with soft, green stems) also live in the jungle understory.

All rain-forest plants are perennials. With so little energy provided by the understory's dim light, annuals would have difficulty growing from seeds and completing reproduction in a year or less. Annual species are best adapted to sunny habitats where the plants can quickly complete their life cycles.

To collect as much light as possible when so little is available, leaves in the understory tend to be broader and thinner than those on the canopy trees. Protected from strong winds that frequently batter the canopy, understory plants freely

expand their leaves and open delicate flowers in an environment where temperatures, although cooler than in the canopy,
vary only slightly between day and night.

Leaf Stickers

High humidity in the understory encourages growth of algae,
lichens, and mosses on the surface of other plants, including
leaves. Such small plants, sticking to leaves, are called *epiphylls*
(*epi*, upon; *phyll*, leaf). Epiphylls obviously reduce a leaf's

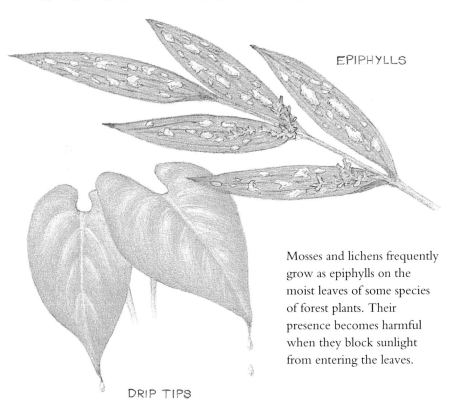

EPIPHYLLS

Mosses and lichens frequently
grow as epiphylls on the
moist leaves of some species
of forest plants. Their
presence becomes harmful
when they block sunlight
from entering the leaves.

DRIP TIPS

To prevent the growth of epiphylls, many rain-forest plants bear leaves that
hang in drooping positions, have waxy surfaces that shed water, and end in
drip tips that channel rain off the blades.

ability to collect light for photosynthesis and, if possible, should be discouraged. To do so, many leaves are smooth and slick, and hang in drooping positions to drain water from their surfaces.

Leaf blades frequently have pointed drip tips that channel rain to the soil and help keep the blades dry. A wet leaf surface encourages growth of epiphylls and reflects too much precious sunlight. It also slows the rate of transpiration, a process needed for water and mineral circulation throughout the plant.

A Surprising Shortage of Soil Nutrients

A strange fact about jungles is that these most densely vege-tated places on earth exist on some of the most nutrient-poor soils. In northern deciduous forests, half the minerals are located in the soil and the other half in the plants. In contrast, most of a rain forest's minerals are deposited in the vegetation, with only a small fraction present in the soil. Some of the plants' mineral reserves are recycled back to the soil when leaves and other plant parts drop to the forest floor.

If you were standing in a rain forest, you would be aware of the continuous shower of dead leaves, blossoms, and twigs bearing their stored minerals for a brief stay on the ground. In the understory's hot, humid atmosphere and under attack by ants, termites, soil fungi, and bacteria, the litter of plant parts soon decomposes. Fallen leaves barely look like leaves after two to three weeks. And in six weeks they are completely decayed. For comparison, in a temperate-zone deciduous forest it takes about one year for leaves to rot, and seven to ten years for needles to decay fully in a northern pine forest.

When minerals in the rain forest's organic litter are washed into the soil by tropical showers, they are quickly sucked up by spreading mats of interwoven, shallow roots. Fine branch roots grow upward and attach to leaves as they rot, there to capture released nutrients even before they have a chance to sink into the ground.

Certain soil fungi assist many of the forest trees in their quest for nutrients, especially the important minerals phosphorus and nitrogen. Thin fungal threads wrap around the tree roots and harmlessly penetrate them. The threads also spread into the nearby soil and leaf litter, drawing in minerals and directly channeling them to the roots. The trees, in turn, provide their fungal helpers with sugary foods, produced in the leaves by photosynthesis.

Biologists call fungi and roots that live in this type of relationship *mycorrhizae* (my-co-*ry*-zay), meaning fungus-root. Mycorrhizae play important roles in the nutrient cycles of many plant communities, but are essential to the life of the tropical rain forests.

Plants and Animals Working Together

The union between fungi and trees is one of many associations that occur in nature between completely different organisms. In each union, or *symbiosis* (sim-by-*o*-sis), both partners benefit from each other's presence. The scientific word means living together.

The survival of a jungle depends to a large extent on the success of its many symbioses. These include fascinating relationships that have developed between plants and animals over the millions of years it took for the rain forests to evolve.

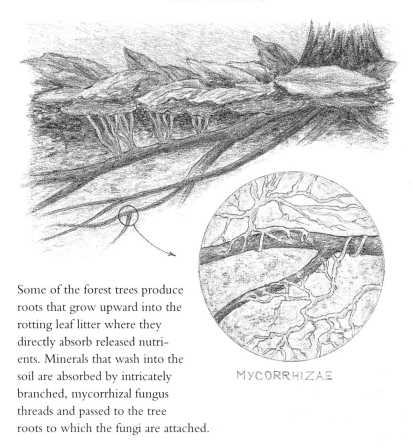

Some of the forest trees produce roots that grow upward into the rotting leaf litter where they directly absorb released nutrients. Minerals that wash into the soil are absorbed by intricately branched, mycorrhizal fungus threads and passed to the tree roots to which the fungi are attached.

MYCORRHIZAE

Jungle animals, large and small, assist in the plants' well-being in a number of ways. Some plants are guarded by ants that chase away leaf-eating animals. As a reward for their efforts, the ants are given comfortable nesting sites in hollow spaces within the plants they protect, plus a diet of specially prepared foods.

While larger animals eat ripe fruits and seeds for nourishment, they also scatter some of the seeds, allowing them to germinate. Many seeds are dispersed throughout the forest when they pass, unharmed, through animals' digestive tracts

71

and are dropped in distant places. Such seed dispersal is essential to the survival of plant species. When seedlings try to compete with their parents by growing close to them, they are likely to lose the fight for water, nutrients, light, and growing space.

Seeds produced in the airy heights of the forest canopy are frequently small and lightweight for easy wind dispersal. Some have attached wings or parachutes to help them fly long distances. However, in the still air of the understory, wind dispersal of seeds is out of the question. So many plants in the lower forest levels must rely on animal dispersal.

Others simply form unusually large seeds that drop directly to the ground. With plenty of stored food in the seeds to rely on, the seedlings have a chance of surviving alongside their parents. The nutritious contents of large seeds make up for poor photosynthesis in the dim light of the forest floor and mineral shortages in the soil.

Many types of insects, birds, and bats play vital roles in the dispersal of pollen. Sometimes a single insect species may be the only type of animal that can carry a plant's pollen from flower to flower and help it reproduce. In turn, that plant species may be the only suitable source of food for the insect. Such dependencies tread the fine line of survival—if one partner becomes extinct, so does the other.

Rain-forest plants produce flowers at irregular intervals. By some mysterious process, all the individuals of one species in an area will flower at the same time. Members of the same species flower in other parts of the forest on different occasions.

Mass flowering in small areas improves the chances of successful cross-pollination between plants. And as a result of such a blooming behavior, large numbers of fruits (derived from flowers) eventually ripen at the same time. By swamping the animals with food, at least a few fruits escape their attention. Consequently, the seeds from these fruits have an opportunity to germinate into new plants.

The Jungle Heals Its Wounds

The rain forest is a dynamic, constantly changing place. As the larger trees become top heavy from growth and the weight of attached epiphytes, their shallow root systems are no longer sufficient to hold them in place. So here and there in the forest a tree crashes to the ground, flattening other plants in its path and opening a gap in the forest structure into which full sunlight streams to the forest floor, perhaps for the first time in hundreds of years.

The death and decay of a fallen tree, its epiphytes, and attached vines releases large quantities of nutrients into the soil. These minerals stimulate the rapid growth of other plants waiting to occupy the newly created environment inside the gap. Shade-loving understory plants that survived the tree's downfall, not accustomed to bright sun and high temperatures, also die and contribute to the enriched nutrient supply.

The natural re-vegetation of a gap occurs in an orderly sequence, or *succession*. It begins with the germination of seeds that arrived after the gap was opened and others that had survived being eaten or rotting on the dark forest floor.

In the intact forest, many seeds are unable to germinate because of the lack of red light reaching the ground. This is similar to the regulation of seed germination in a deciduous forest where canopy leaves also filter out red wavelengths.

73

When an old giant falls and takes some of the surrounding trees with it, a gap is opened in the forest structure (top). The gap quickly fills with sun-loving pioneer species (middle). As the young forest trees grow and spread their branches, the temporary sun-lovers die in the increasing shade. In time, the structure of the rain forest is restored (bottom).

74

There, the seeds must wait to germinate until the canopy trees
have lost their leaves; in the jungle they must wait until a gap
is formed.

The succession continues with the rapid growth of all the
sun-loving plants. These include saplings of canopy and forest
giant species that were undamaged by the fall of the gap-
forming tree. Such saplings had grown at a rate of only about
1 inch (2.5 cm) per year in the understory's shade. But in the
gap's full sun, 40 inches (1 meter) of growth per year is not
unusual.

Pioneer species occupy gaps during the first years of the
succession. These plants specialize in quickly filling open areas
as they occur throughout the forest. Pioneer species include
many types of small, short-lived trees that thrive in the
warmth, bright light, and nutrient-rich soils that the openings
provide.

When the pioneer vegetation casts a deepening shadow on
the ground, shade-tolerant species begin to return to the area.
And as longer-lived canopy and forest giant trees mature, they,
in turn, overshadow the pioneer forest, spelling its doom.
Succession ends usually within 100 years after the gap was
opened. At this point the cycle of regeneration is complete—
the canopy and understory match the surrounding,
undisturbed forest, and no trace of pioneer species remains.

Surviving intense competition and overcrowding in a
jungle is as much a challenge as living through the changing
seasons farther north or south of the equator. The many
microenvironments in the rain forest are occupied by different
plant species, each with survival methods as artfully devised by
nature as those of plants in other habitats throughout the
world. These methods include:

❧ The ability of understory species to photosynthesize in dim light, and to keep their leaves free of epiphylls.

❧ The ability of lianas and other vines to escape from the dark forest floor by carrying leaves on long, thin, clinging stems that rapidly grow into the brighter light of the treetops.

❧ The special trick of epiphytes to spend a lifetime clinging to the high branches and trunks of trees, and there to collect water and minerals by unusual means.

❧ The ability of canopy trees to start life in the understory, grow slowly in the shade for many years, but be ready for rapid growth when gaps occur.

❧ Formation of seeds especially designed for wind dispersal (in the canopy) or animal dispersal, and germination in either the understory or in gaps when the opportunity arises.

❧ Entering into a symbiosis: getting a fungus or animal to help growth or reproduction by supplying it with food and other benefits.

In the intricate organization of a tropical rain forest, plants create microenvironments for their fellow plants to live in, and the vegetation interacts with an abundance of animals and other life forms. All depend on the soil's meager offerings and a climate that, although unchanged from year to year, is known for its excessive heat, rainfall, and humidity.

Because the amount of water vapor transpired into the atmosphere above jungles is so great, scientists believe that these forests affect the earth's climate more profoundly than any other ecosystem does. This is why scientists also believe that rain forests indirectly affect all of our lives.

THE DESERT'S OPEN SPACES

CHAPTER 4

SURVIVING SUMMERS
IN A DESERT

Arid Places

It takes only a few hours to fly from Costa Rica in Central America, with its lush rain forests, to Phoenix, Arizona, located in one of North America's great deserts. So anyone who takes such a journey would see and feel, in less than a day, the extreme differences between two of the earth's major ecosystems.

After the jungle's dense, enclosing plant life and deep shadows, the traveler would be dazzled by the desert's wide open skies and distant, sparsely vegetated landscapes. And away from the sticky tropical humidity, the crisp, dry desert air would feel comfortable.

Deserts typically receive 10 inches (25 cm) or less of rain each year. Most of it falls in winter months when temperatures are cool, sometimes dropping below freezing. In the northern hemisphere, desert winters usually last from November to January or February; in the southern hemisphere, from June to August. Of the four seasons, summers are least favorable for most desert plants. With intense heat, parched soils, and blinding light, summer months challenge the survival instincts of every plant and animal daring to make its home in the world's arid places.

Deserts are often thought of as vast, sandy wastes without a plant in sight—similar to pictures of parts of Africa's Sahara Desert. But this type of landscape is not typical of most deserts, where vegetation may be meager but quite visible. And instead of shifting sand dunes, the soil surface is baked into a hard crust. On the flat desert floor, in the shadow of rocks, or on the sides of hills and mountains, a host of microenvironments supports many different plant species, each with its special methods for staying alive.

Summer's Endurance Test

Desert plants survive summer by following one of several strategies. Large perennials, with ways to prevent water loss from their stems and leaves, resist drought by slowing growth. Because young, growing plant parts are most easily damaged by heat and water shortages, the plants wait for the coolest and wettest seasons before allowing tender shoot tips to emerge. Some plants lose all their leaves and enter full dormancy in summer. Perennials growing from bulbs, rhizomes, or tubers let their above-ground stems and leaves die back, and sleep through the hottest months, safely underground.

Deserts are home to large numbers of annual species. Desert annuals avoid heat and drought by existing simply as seeds for most of the year. Some seeds remain dormant for many years, waiting for the right amount of winter rains to ensure both seed germination and survival of the new crop of plants. Because annuals can grow only in well-moistened soil, their

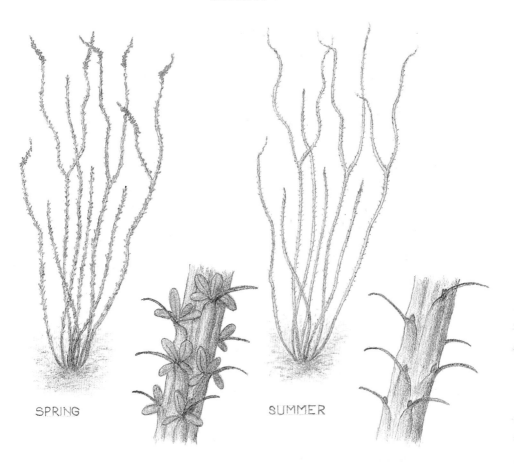

SPRING   SUMMER

In spring, the tall, slender stems of ocotillo bear clusters of small leaves.
Photosynthesis builds up the plant's food reserves to last through the long
summer when the leaves drop off. Sharp spines protect the ocotillo's stems.

life spans are limited to a brief two to four months in winter
and early spring. Some complete their seed germination,
growth, flowering, and seed production in as little as six weeks.

Because seeds themselves are extremely dry, they have no
problem staying alive during several years of drought. And they
are even unharmed by high soil temperatures, sometimes
reaching 175°F (80°C) in midsummer days in the desert.

81

Baked Soils and Rushing Rivers

High temperatures and the shortage of water in desert soils discourage fungal and bacterial activity, necessary for the decay of dead plant parts. Because it may take years to decompose the small amount of leaf litter that the plants produce, desert soils are very poor in this organic matter.

Desert soils are often rich in minerals—in some cases, too rich. Minerals slowly released from eroding rocks are not washed deeply into the ground, as they are in places having frequent rains. And in some desert areas, underground minerals are brought to the soil surface in water moving upward when it evaporates into the atmosphere during hot weather. In the resulting salty soils very few plants are able to grow.

The hard-packed crust of the desert pavement, as the soil surface is called, allows only shallow penetration of water even in the heaviest rains. With only a small portion of the rain soaking into the soil, most of it runs off the pavement, down slopes, and into dry river beds, or washes.

Shortly after a heavy downpour, so much water drains into the washes that they become raging torrents. Churning rocks and gravel destroy all plants in the water's path, except firmly rooted trees.

Some of the trees that typically grow in washes have seeds with extra-hard coats. These require the grinding action of gravel moving down the flooded wash to make holes in the coats. Only then can the seeds absorb water and germinate. Smoke tree and ironwood seeds must be swept 150 to 300 feet (45–90 meters)—but not more—down a wash for the necessary amount of scratching to occur. Beyond 300 feet, the seeds are pulverized.

Some desert trees, such as ironwood, live in riverbeds, called washes. There the hard-coated seeds have a chance to germinate. 1. The seeds drop off the tree and wait for a heavy rain to flood the wash. 2. Carried down the wash with sand and gravel, the seeds get their coats worn away, allowing water to enter the seeds. 3. Seeds that settle after a partial grinding germinate in the wash. 4. Seeds washed beyond that point are pulverized. Such a system places young trees a distance from their parents where they have less competition for water and nutrients.

Some Cactus Secrets

To capture the small amount of life-giving water that sinks into the desert pavement, many perennial species, including cacti, have wide-spreading, shallow root systems. Desert annuals also have shallow roots—sufficient to support the plants until they make seeds before dying in the summer sun.

To survive for long periods without rain, a cactus relies on water stored in its fleshy stems. When rain first reaches the dry cactus roots, they quickly form new, living tissues that work

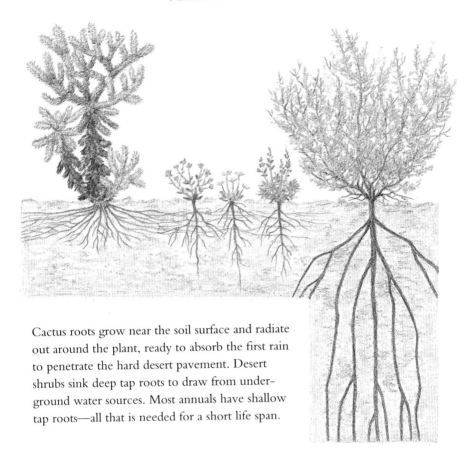

Cactus roots grow near the soil surface and radiate out around the plant, ready to absorb the first rain to penetrate the hard desert pavement. Desert shrubs sink deep tap roots to draw from underground water sources. Most annuals have shallow tap roots—all that is needed for a short life span.

fast to draw moisture from the soil and stuff it into soft stem tissues. This will become the only source of liquid for the plant's metabolism until the next rain, perhaps in two, three, or more years.

Several cacti, including the giant saguaro, have stems with parallel ridges on their surface. The ridges partly shade the stem, helping to keep it cool. The pleated stems also expand quickly when water is stored and contract slowly as the moisture is used. Specimens of saguaro cactus, 30 feet (almost

10 meters) tall, contain hundreds of gallons of liquid. Some of the largest plants are estimated to be more than 200 years old.

Of course, it is important that a plant such as a cactus not waste water, especially by evaporation from its body. A thick, waxy coat (a cuticle) on its above-ground parts is the first line of defense against drying. And cacti lack normal types of leaves that easily lose water by transpiration; photosynthesis occurs in the green stems where few stomata are present.

GIANT SAGUARO

Ridges on the stems of the giant saguaro cactus provide shade to keep the plant cool. They also increase the stem's surface area to promote maximum photosynthesis. Pleated stems can expand without splitting when stored water fills their tissues. Spines protect the succulent stems against thirsty animals.

Cacti with flat stems point their branches in different directions. Throughout the day, only a portion of the branches directly face the sun; the others shade themselves and so stay cool.

Several cactus species have flattened stems. When branches grow, they are positioned in various directions. This way, different portions of the plant always stay cool by facing away from the sun as it moves across the sky.

As cacti evolved, regular leaves changed into more useful, prickly spines. Although a cactus may be desirable to thirsty animals, few creatures will bite into the plant's spine-covered surface. To survive in deserts and other habitats, plants need to defend themselves against animals in search of food and moisture. They protect themselves either with special structures, such as spines or thorns, or with bad-tasting (even poisonous) chemicals stored in leaves and stems.

Such survival methods are especially important for desert perennials that grow slowly in the harsh environment. It takes a long time for them to replace body parts eaten by hungry animals. And it is understandable that, with water being so

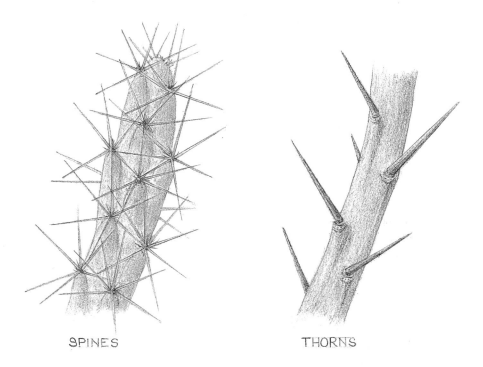

SPINES   THORNS

Some desert plants have spines or thorns that discourage animals from eating them. Cactus spines evolved from leaves and most often occur in clusters. Thorns are short, pointed, woody branches produced by several tree and shrub species.

difficult to get, desert plants are reluctant to share it with thirsty predators.

Tapping Underground Water Supplies

Deep below the desert's surface, pockets of water collect, ready for use by plants with roots long enough to reach them. Most trees and other large desert perennials grow *tap roots* reaching 90 feet (30 meters) or more into the ground. For such plants, the short period immediately following seed germination is the most uncertain time of their lives. The seedlings' energy must be put mostly into quickly sinking tap roots into the soil, with the hope of reaching an underground water source. Only if the roots are successful can shoot systems begin to develop.

Because deep water sources don't evaporate from the soil in the summer's heat, they offer a year-round supply of moisture to plants able to reach them. Even so, to conserve water, some desert trees and shrubs with deep tap roots still shed their leaves in the hottest months. They simply rely on green stems for continued photosynthesis.

Protection Against Bright Light

In the openness of the desert, plants receive plentiful sunlight. Often it is more than they need for making their required food supplies. If the light is too bright, a plant's photosynthetic system may be harmed. To prevent this, the leaves of some species rotate on their stalks to avoid facing directly into the mid-day sun. Others reflect light from dense mats of surface hairs. On some cactus stems, masses of light-colored spines serve a dual purpose—they protect against animals and reflect as much as three-fourths of the sunlight striking them.

Competition for Water

With so much light available in the desert, the plants never have to struggle to reach it as they do in a rain forest. But because water is limited, larger plants do well to avoid crowding each other and wasting energy on competition for moisture in the soil. The number of perennials the land can support is dictated by the average yearly rainfall. It is up to the plants to control the spacing between them.

In places such as Arizona's Sonoran Desert, a fairly regular rainfall pattern includes both winter and summer rains. As a result, the perennials grow close to each other. On the other hand, in California's Mojave Desert, most of the year's rain falls in winter—if it rains at all. The resulting vegetation is sparse, the perennials well spaced.

Members of some species, such as the creosote bush, are so evenly separated from one another that they look as though they were planted by someone with a measuring stick. Perennials that try to compete with the creosote bush, by filling in the spaces between them, have little chance of growing beyond the seedling stage—and that includes other creosote bushes. Apparently, the spreading roots from established plants draw so much water from the soil that intruders find it difficult to survive. Only annuals with shallow roots and small water needs grow successfully among the bushes.

Chemical Warfare

Some researchers say that creosote bush may also stop other perennials from growing near it by giving off a deadly chemical from its roots. There is strong evidence that other plant species in dry regions do use a type of chemical warfare to reduce competition for water. For example, the fallen leaves of

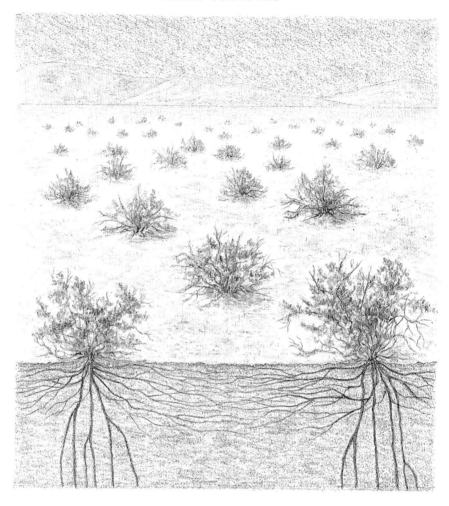

Creosote bushes space themselves across the desert floor. Their spreading roots draw so much water from the surface soil that other perennials are unable to survive near them. Shallow-rooted annuals have a better chance of living among the creosote bushes.

brittlebush, a native of the American Southwest, collect around the shrub's base. Because the leaves release a substance that is toxic to the seedlings of most species, the brittlebush is able to keep a clear patch of ground around itself. This method, *allelopathy* (al-*el*-o-pathy), only works in places that have low rainfall; heavy rains simply wash the effective chemicals out of the soil.

Brittle bush reduces competition for water by creating a no-growth zone around itself. A chemical released from the brittle bush's dead leaves, scattered below the plant, prevents the growth of seedlings.

Regulating Seed Germination

Most desert annuals have no problem living in dense plant populations. When there is plenty of water, the plants grow to their maximum size. In times of lesser rainfall, they are smaller but still large enough to bear flowers and complete their life cycles to seed production. Being so flexible gives the annuals an added advantage for surviving.

Because the amount of rain varies from one year to the next, a show of annuals in the desert cannot be counted on. The seeds of some species of desert annuals contain chemicals that actually prevent them from germinating. Only when enough rain falls to wash these substances out of the seeds do they break dormancy. And, of course, when the rainfall is heavy the soil is sufficiently well soaked for the seedlings and adult plants to survive.

In deserts that have two rainy seasons, the species of annuals that germinate after the winter rains are different from those germinating in summer. How does each set of seeds recognize the season when it is supposed to grow? Experiments have shown that "winter seeds" only germinate in cool temperatures, about 60°F (15°C), after a heavy rain. "Summer seeds" only respond to moisture and warmth, about 85°F (32°C).

Even under the best conditions, not all the seeds in the desert soil will germinate at the same time. Frost may kill the seedlings in winter, or rapid drying may kill them in summer. So a portion of the seed bank remains dormant, waiting for other opportunities to grow. Always having reserve seeds gives species a way to survive the desert's harsh and unpredictable climate.

Seasons of Growth

Winter is the principal season for renewed growth in the desert. After a series of early winter rains, the desert's complexion is completely transformed. The drab, brown desolation of the desert floor changes to fresh green as young grasses and other annuals emerge from long-dormant seeds. Shriveled shrubs and trees explode with new foliage. Shrunken cacti and other succulents swell with the welcome water. And in rocky crevices, ferns and mosses make a surprise appearance—plants you would hardly expect to find in a desert.

In winter, night temperatures are chilly; snow occasionally covers the ground. So, in their first stages of new growth, the desert annuals make slow progress. They keep their leaves pressed flat against the soil to capture its warmth in the day-time sun. Roots keep digging deeper to avoid drying in the surface soil. And as small as the plants may be, each carefully measures the length of the passing days, waiting for a precise time to begin flowering.

With the approach of spring, the plants have put out several sets of leaves that they elevate into the warm air on fast-growing stems. Flower production is stimulated by the season's longer days and shorter nights—a response to the environment called *photoperiodism*.

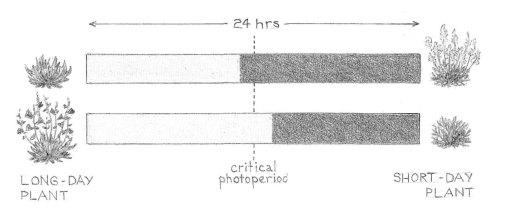

24 hrs

critical
photoperiod

LONG-DAY
PLANT

SHORT-DAY
PLANT

Photoperiodism—the effect of day and night length on plants—regulates flower production in many species. Some species (long-day plants) only flower when the seasonal day length exceeds a certain critical photoperiod. These are the plants that flower in spring and early summer, as do most desert annuals. Short-day plants flower in response to day lengths that are less than their critical photoperiod, such as in later summer and early fall when day lengths become shorter. The actual length of the critical photoperiod is different for each species.

Many species of desert annuals remain small and close to the ground, eventually forming one or two flowers on tiny stems bearing a few small leaves. One scientist called them "belly plants" because you have to lie on your stomach to look at them closely. Belly plants are only some of the many species of desert plants classed as *ephemerals* (e-*fem*-er-als), meaning they have only a short existence.

"BELLY PLANTS"

Among the desert's vast array of annuals, some species produce only a few small leaves and flowers on tiny stems. If you lie on your stomach to study them close up, "belly plants" are miniature masterpieces. Here a pencil point is given for scale.

In "good years" with the right mix of winter rains and favorable temperatures, the ephemerals reach their peak of growth by the middle of spring. The desert, so bleak for most of the year, is carpeted with flowers of every imaginable color in one of nature's most spectacular displays. And the air is alive with insects that, in a feeding frenzy, ensure a bountiful seed crop by carrying pollen from flower to flower. When mice and other desert creatures emerge from their winter dens, a world filled with luscious food is laid out before them.

But such abundance is short-lived. By the end of spring, as higher daytime temperatures evaporate water from the soil, the annual plants hurry to complete their reproduction. Then they slowly disappear from the landscape when winds scatter their dried remains. And seeds, for which the little plants so briefly existed, are spread across the land to wait their turn to start another cycle of life.

For perennials, winter and spring months are also the time of greatest growth. Shallow roots explore the soil in ever-widening circles around the plants. Stems add a little to their length, although growth is generally slow from year to year. New leaves convert the sun's energy into food that is carefully stored for summer use—when photosynthesis may be slowed by heat and lesser water supplies.

A Desert Dilemma

For photosynthesis to occur, carbon dioxide (CO_2) must enter the plant from the atmosphere through open stomata. However, these same tiny pores allow water vapor to escape from the plant by the process of transpiration. The warmer the temperature, the faster water is lost. So on a hot summer day, a desert perennial may transpire faster than it is able to draw water from the soil, resulting in wilting, even death.

Most desert perennials bear small leaves, which absorb less heat and so reduce their rates of transpiration. The leaves also slow down the loss of water by simply closing the stomata. But that prevents the entry of carbon dioxide and stops photosynthesis. Because photosynthesis and saving water are equally important, what is a plant going to do? Succulents have solved the problem with a rather clever system.

Succulents open their stomata at night and store CO_2 in their leaves until the next day. At sunrise, as the temperature increases, the pores close and the stored CO_2 is used for photosynthesis. Because night temperatures in summer are much cooler than daytime temperatures, only a small amount of water is lost when stomata open in the dark.

On the Sides of Desert Mountains

The succulents' system for saving water enables them to occupy the driest sites in the world's deserts. However, their growth on the dry sides of desert mountains is limited by the low temperatures that come with elevation. At a certain level on each mountain, winter frosts reach down to kill the succulents by turning their stored water into ice.

So if you were to climb a mountain slope from the low desert, you would find a point where succulents disappear and are replaced by perennial species able to survive winter's freezing temperatures. Some, including the strange-looking Joshua tree in California's Mojave Desert, only grow at these higher elevations. Living more than 2500 feet (750 meters) above sea level, Joshua trees survive the whole range of desert

conditions: from summer's scorching heat to the frost of winter; from times of plentiful water to long seasons of drought.
Few plants in the world could claim to be so versatile.

If you look across a desert at high noon on a summer day,
you may wonder how any plant could live in such a place. But

JOSHUA TREE

Deserts are home to many strange-looking plants, including the Joshua tree.
New leaves develop at the tips of thick, branching trunks. Old leaves fold
back to protect the stems and insulate them against winter's cold and the
intense sun in summer.

97

plants do grow there, using some of nature's best-crafted methods for survival. These include:

❀ Avoiding drought. For most of the time, annuals exist as seeds; some perennials retreat into underground bulbs, rhizomes, or tubers. Only when the soil is well soaked do these species form whole plants, completing short life cycles at the wettest time of the year.

❀ Resisting drought by having small leaves, or no leaves at all; photosynthesis in green stems; succulent stems and leaves, capable of storing large amounts of water.

❀ Having spreading, shallow roots to absorb rain quickly as it sinks into the topsoil, or deep tap roots reaching underground pockets of water.

❀ Spacing themselves if they are large perennials to avoid competition for water, and defending themselves against thirsty animals.

❀ Germinating seeds only under special conditions that ensure seedling growth in well-moistened soils.

❀ Opening stomata at night (for succulents) to reduce water loss, while still performing normal photosynthesis.

Perhaps one day scientists will plan to colonize another planet, such as Mars. When selecting plant species to take on the mission, they will probably look for those best able to survive drought, intense light, and extreme shifts in temperature. For such a task, the earth's desert inhabitants are made to order, having been thoroughly tested in nature's laboratory for millions of years.

A KELP FOREST

CHAPTER 5

SURVIVING IN WATER

Ancient Ties

Scientists believe that life on earth began under water and
continued to evolve there for billions of years. When
primitive plants and animals finally emerged from seas and
lakes, and developed a new way of living on land, they never
lost their ties to their ancient home. Even plants adapted to
the dryness of deserts only survive when and where water is
available. Annuals grow after heavy rains; deep-rooted shrubs
and trees live where underground water can be tapped.

At a more basic level, the complex metabolism of living
organisms only occurs when chemical substances react in
water in cells and tissues. So the very processes that keep
plants and animals alive depend on the presence of the liquid
in which life itself began.

The Good and Bad Sides of Life Under Water

Life probably began under water because of the many advantages such a place has to offer. For example, water warms up
and loses heat slowly compared with air. So submerged plants

are safe from the rapid temperature changes that land plants experience between day and night. And because water reflects light and acts as a filter, plants below the surface are protected against excessively bright sunlight. The harmful effects of ultraviolet rays are only slightly reduced.

An aquatic plant's stems and leaves directly absorb water and minerals from their surroundings. Buried roots simply hold the plant in place to stop it from floating away. By comparison, land plants must have elaborate root systems for drawing moisture and nutrients from the soil. And they must have methods (including transpiration) for transporting water to the above-ground shoots.

On the other hand, always being under water has its problems. Oxygen gas, needed by both plants and animals, is never as freely available in water as it is in the atmosphere. Only when water is in rapid motion do large amounts of oxygen get churned into it. In nature, aquatic plant species with high oxygen requirements can only grow in such places as fast-moving streams or coastal regions bathed by ocean surf. In an aquarium, plants grow best with plenty of bubbles provided by an air pump.

The filtering effect of water limits the depth to which light can penetrate and, therefore, the places where plants can grow. Water also alters the spectrum of light, quickly absorbing the sun's red wavelengths used by most plants for photosynthesis. Judging from the colors you see in underwater photographs, you could guess that only the blue-green part of the visible spectrum penetrates deeply into water.

Staying Close to the Surface

Because light intensity and the amount of red decrease with depth, most aquatic plants live as close to the water's surface as possible. *Phytoplankton* (fito-*plank*-ton) are the many thousands of species of microscopic, lightweight algae that float in oceans, lakes, and ponds. These tiny plants move toward or away from the surface, into bright or dim light, depending on their daily photosynthetic needs.

Phytoplankton are a rich source of food for aquatic animals. Several species of freshwater phytoplankton are shown here as they appear through a high-power microscope.

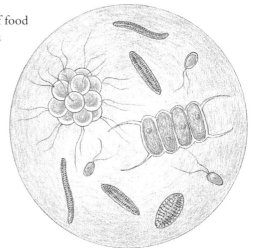

Phytoplankton are some of the most important plants on earth because they are the beginning of the animal food chain. Phytoplankton make food by photosynthesis; small animals eat the phytoplankton; larger animals, including fish, eat the small ones, and so on. The food chain continues when land animals, including humans, eat fish as part of a diet that includes vegetables and fruits.

Phytoplankton float freely in the water. But larger aquatic plants are generally anchored in one place, using gas-filled, hollow stems or flotation bladders to reach toward the surface. For example, the hollow stems of elodea, a popular aquarium plant, float its leaves high in the water. And bladders are a feature of many seaweeds attached to rocks along the world's coastlines.

On the ocean surface the largest seaweeds, the giant kelps, float leaflike blades with bladders that may grow to several inches in diameter. Blades and bladders are attached to stems, or stipes, that may grow over 100 feet (30 meters) long. Supported by the water's buoyancy, kelp plants stand upright without the woody trunk that tall trees are forced to build on land. To prevent the kelp plants from being washed ashore by heavy tides, rootlike holdfasts securely grasp underwater rocks.

How Red Seaweeds Live in Deep Water

Red algae are another type of seaweed with holdfasts for sticking to rocks. Many species of red algae grow in the dim, blue-green light of deep water. Although land plants and other seaweeds could not survive in such conditions, red algae collect the small amount of available light with a unique red pigment, *phycoerythrin* (fico-er-*rith*-rin); *phyco* means seaweed, *erythro* means red.

Phycoerythrin traps the energy in blue-green rays and passes it to the plants' chlorophyll. Chlorophyll, in turn, transfers the energy into photosynthesis. Land plants never contain phycoerythrin. Although you may have seen trees with red leaves, they are colored by anthocyanin, which is not involved in photosynthesis.

By occupying a habitat where other plants are unable to live, red algae have few competitors for growing space. In

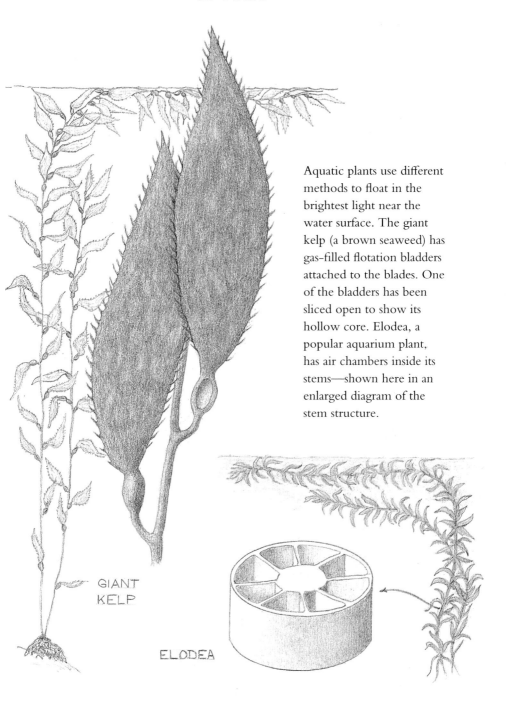

Aquatic plants use different methods to float in the brightest light near the water surface. The giant kelp (a brown seaweed) has gas-filled flotation bladders attached to the blades. One of the bladders has been sliced open to show its hollow core. Elodea, a popular aquarium plant, has air chambers inside its stems—shown here in an enlarged diagram of the stem structure.

GIANT
KELP

ELODEA

105

RED ALGAE

The brilliant color of red algae comes from phycoerythrin, a pigment that collects the filtered blue-green light under water and uses it for photosynthesis. The hundreds of species of red algae grow in a variety of shapes and sizes.

106

their underwater retreats, they avoid being beaten by waves that crash against seaweeds on rocks near the ocean surface. And by living below the level of the lowest tides, deep-growing red algae are never exposed to the air and hot sun.

Surviving Dry Periods

Anyone who has been near the ocean is familiar with brown and green seaweeds that droop from rocks at low tide. And you probably know how easy it is to slip on these plants. Their bodies are coated with *mucilage*, a slick, thick slime. Mucilage slows water loss from the seaweeds when they are exposed to the air. As the mucilage coat dries, it tightens around the plants, giving them increased protection, at least for the few hours while the tide is out. However, mucilage cannot protect against prolonged exposure to the sun.

A longer-lasting protective coat surrounds the cells of microscopic algae inhabiting temporary pools. The coat, or *gelatinous sheath*, swells when soaked in water, allowing the cells to perform their normal life functions. But when a pool

Gelatinous sheaths. Algae that live in small ponds face the danger of the pools drying up in summer. Some such plants coat their cells with gelatinous sheaths. The algae seen here through a powerful microscope are one-celled but divide into two-, then four-celled colonies. The colonies, held together by their sheaths, eventually break apart and the individual cells repeat the colony-forming process.

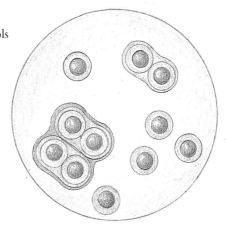

dries up and the algae come to rest on the muddy bottom, their sheaths slowly shrink and harden into a coat that protects the cells inside against complete drying. The cells enter dormancy and stay that way until the pool is reflooded, perhaps after several years.

The Water Lilies' Air-Conditioning System

Water lilies live in two different environments. Their leaves float on the water's surface, in contact with the atmosphere. Stems and roots lie buried in poorly aerated mud at the bottom of a pond. Air enters the leaves through stomata

Water-lily leaves absorb air through tiny stomata in their upper surface. Underwater roots and stems are supplied with oxygen that passes from the leaves through an elaborate system of internal ducts.

located in the upper leaf surface, then moves to submerged parts through large, internal ducts connecting the leaves, stems, and roots.

To prevent the stomatal pores from becoming clogged with settling dust, the leaves are coated with a smooth, waxy cuticle layer. When water splashes onto the cuticle, it forms rounded droplets that roll across the surface and off the sides, carrying dust with them.

Knees That Breathe

Roots must be constantly supplied with oxygen to stay healthy. Most plants with submerged roots and above-water leaves solve the problem of oxygen transport with air-conditioning systems similar to the water lilies'. But bald cypress, an inhabitant of swamps in eastern North America, has a more direct method. From the roots of the bald cypress, woody extensions, called knees, grow upward into the air. Oxygen passes through tiny holes at the top of each knee and moves quickly through the porous core. To keep above rising water levels, the root extensions simply grow longer, until they resemble periscopes poking out of the water around the base of each tree.

The special root feature of the bald cypress helps it survive in a swampy habitat. But few plants produce knees. If the roots of most trees are submerged in water for extended periods, the trees die from suffocation.

Another problem occurs when soils become waterlogged (saturated with water). The water displaces oxygen that normally occupies spaces between the soil particles. A shortage of oxygen in the soil encourages growth of *anaerobic* (an-er-*robe-*

Bald cypress knees. In a Florida swamp, the roots of bald cypress trees get oxygen from the atmosphere through porous knees that project above the water surface. Spanish moss (a type of bromeliad, and an epiphyte) hangs from the tree branches.

ic) bacteria—ones able to live in the absence of oxygen. These organisms are especially destructive to roots, but their presence is frequently easy to detect. When they decompose organic matter, they give off gases that smell like rotten eggs.

Features of Underwater Leaves

When an aquatic plant's leaves, stems, and roots are all submerged, the plants must draw carbon dioxide (CO_2) and oxygen directly from the water. Because CO_2 readily dissolves in water, it is plentiful and easily absorbed. But oxygen, which occurs as tiny bubbles, is often in short supply, especially in still water. To make up for the oxygen shortage and to come into contact with as much of the gas as possible, underwater plants must have large surface areas. The evolution of finely divided, feathery leaves by many aquatic species fulfilled such a need. The popular aquarium plant, myriophyllum, is a good example.

Feathery leaves also let flowing water pass over them without harm. Large, flat leaf blades easily tear but, above water, are better for collecting light for photosynthesis. Water buttercup solves both problems by making two types of leaves: the feathery type underwater, broad leaf blades poking into the air.

Adaptations for Living in Moving Water

Water is much heavier than air. So when water is in motion it strikes objects with a far greater and more damaging impact than wind blowing at the same speed. Plants daring to live in fast-moving streams, small waterfalls, or coastal surf are constantly battered by the moving water.

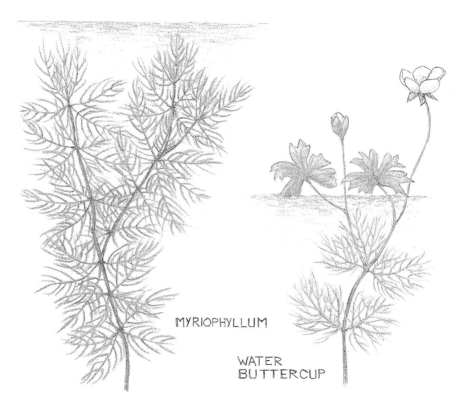

MYRIOPHYLLUM

WATER
BUTTERCUP

The feathery leaves of some aquatic plant species have extra-large surface areas to draw oxygen out of the water and to reduce damage in flowing water. Water buttercup produces both feathery leaves under water and broad blades above the surface.

Although turbulent aquatic habitats are dangerous places to live, they do provide certain benefits. Plenty of oxygen is churned into the water, and dissolved minerals are constantly in circulation. But to do well in such places, the plants must be firmly attached to solid rock, and have streamlined, flexible bodies.

Eelgrass and sea palm are two good examples of species with those features. To remain undamaged by the sea's constant motion, eelgrass has long, narrow leaves that gracefully sway back and forth with each passing wave. Spreading roots attach the plants to rocks. Sea palm (a brown seaweed) is perfectly designed to survive in the heaviest surf. Its short, rubbery, stemlike stipe is topped by tough, flexible blades. Strong holdfasts firmly anchor the plants to solid rock. At high tide, the sea palms show off their endurance. Under

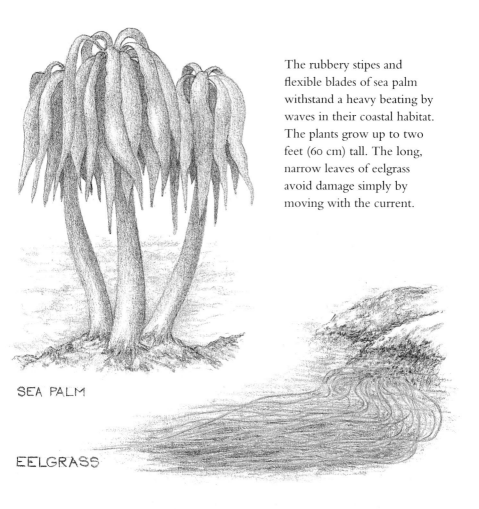

The rubbery stipes and flexible blades of sea palm withstand a heavy beating by waves in their coastal habitat. The plants grow up to two feet (60 cm) tall. The long, narrow leaves of eelgrass avoid damage simply by moving with the current.

SEA PALM

EELGRASS

the impact of each crashing wave, the stipes bend in dancelike rhythm, then quickly spring back to their upright positions.

Fresh Water or Salt Water?

Few aquatic plants are designed to survive the beating taken by sea palms. Most occupy quieter waters where physical strength is rarely tested. Species living in freshwater lakes, ponds, rivers, or streams, and those adapted to the salty waters of oceans and seas are clearly different.

Water preference (fresh or salty, but not both) is based on the inherited structure and function of each plant's cells. A freshwater plant dies from the excess salts in seawater; a seaweed in fresh water gorges itself on so much liquid that its cells burst.

Try testing the reaction of an aquatic plant to a "wrong" environment. Prepare several solutions of table salt at different concentrations, starting with a very weak solution. Divide an aquarium plant into equal pieces and place one section in each solution, plus an extra piece in plain water. During several days, note the appearance of each plant and how long each of them lasts. You could do a similar experiment to compare seaweeds in fresh water and sea water.

Dealing with Salt in a Coastal Marsh

Salt marshes occur in low-lying coastal areas where the vegetation traps sediment that is washed ashore by the tides. The plants' roots, stem and leaf fibers, and the remains of partly decayed plants and animals bind together into *peat*, a soft, springy mat. During high tides, when the marsh floods with sea water, salts soak into the rich, organic soil mixture.

Salt marsh plants survive changing environmental conditions: They are submerged in sea water at high tide, and, during rains, they are washed in fresh water. But they are most cleverly adapted to life in the salty soils. Botanists have studied one of the most common marsh plants—spartina, or cord grass—to learn the secret of its success.

The cells in spartina's roots have unusual membranes that permit water to enter, but filter out most of the salts. Salts managing to cross the membranes are stored in special cellular compartments and squeezed out of glands located in the leaves. On the leaf surfaces, the salts dry into white specks, then are washed back to the soil by rain. Tiny pores in the salt glands also function as openings through which oxygen enters the plant before passing to roots through a system of internal tubes. The pores close when high tides begin to submerge the plants.

Peat Bogs in the Making

In still waters around the margins of freshwater ponds and small lakes, water lilies and other floating plants take root while cattails and reeds crowd into the shallows. These are the pioneer species in a succession that may transform the habitat from a body of open water into a mature forest.

Between these two stages, after the water has filled in but before the trees appear, the area will support a soft, soggy peat bog with its own specially adapted plant species.

Peat begins to form early in the succession when partly decayed organic matter collects among the cattails' tangled roots, rhizomes, and stems. Using this as a fairly solid base, other plants, including sphagnum moss, grow out from the

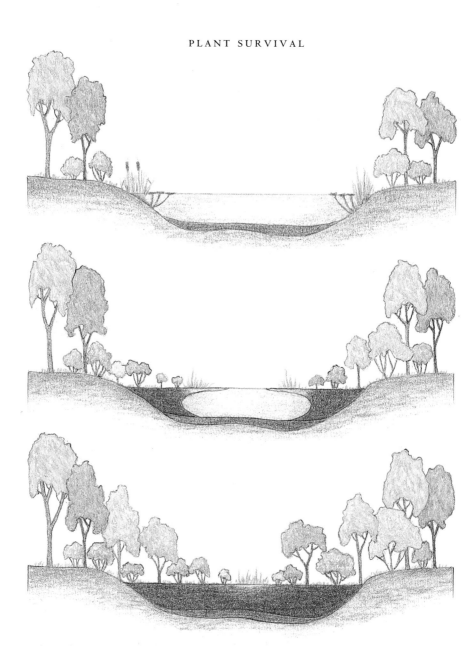

The succession leading to development of a peat bog starts with a pond or small lake and the plants that crowd the shallow water around its shores (top). Sphagnum moss slowly forms a living, floating mat, and thick deposits of dead plants collect under it (middle). The bog is complete when there is no more open water (bottom). Eventually, a forest of trees and shrubs occupies the area where there was once a pool of water.

116

shore. At water's edge, they form a floating mat that slowly advances over the lake surface. Year after year, as dead sphagnum collects under the mat, it turns into thick peat deposits. This base supports an ever-heavier load of vegetation, including low-growing shrubs such as cranberry.

The bog is complete when open water is no longer visible—the water having been absorbed by the peat or evaporated into the atmosphere. Some time later, around the margins of the bog, the first trees germinate from seeds introduced from the surrounding forest. As the trees draw water from the peat, the soil becomes dryer and better

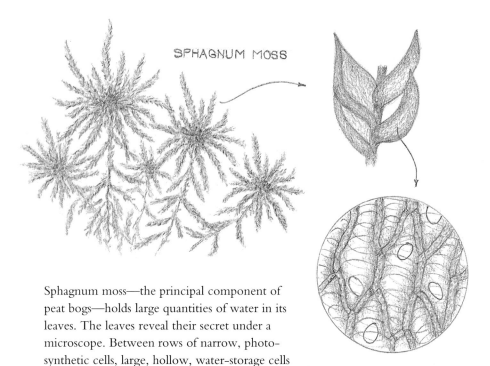

SPHAGNUM MOSS

Sphagnum moss—the principal component of peat bogs—holds large quantities of water in its leaves. The leaves reveal their secret under a microscope. Between rows of narrow, photo-synthetic cells, large, hollow, water-storage cells make up the bulk of the leaf. Water enters through holes in the storage cells' walls.

aerated—conditions that encourage other tree species to move in. The succession reaches its climax when bog plants, no longer able to grow in the dry soil, are replaced by pure forest.

Plants That Eat Animals

Peat bog soils are especially lacking in nitrogen, an essential mineral for the growth of both plants and animals. (It is needed for building protein molecules.) The nitrogen shortage limits the number of plant species that can occupy these habitats. But one group overcame the problem when it evolved a most ingenious adaptation: the plants extract nitrogen from the bodies of captured insects.

The *insectivorous* plants (in-sec-*tiv*-or-us), or insect-eating plants, are flowering plants that make food by normal photosynthesis. What makes them different are leaves that have evolved to lure, trap, and digest insects and other small animals whose bodies contain nitrogen-rich proteins.

They capture insects in several ways. For example, the sundew has leaves with sticky hairs in which the animals become tangled. Pitcher-plant leaves consist of hollow tubes that lure visitors to a deadly pool of liquid at the bottom. Slippery surfaces or thousands of stiff hairs pointing downward prevent the insects from climbing out of the tubes. Venus' flytrap has hinged leaf blades that snap shut on visitors when they brush against tiny trigger hairs.

After capture, the insect bodies are slowly digested by enzymes excreted from special glands, or by bacteria living in a pitcher plant's pool.

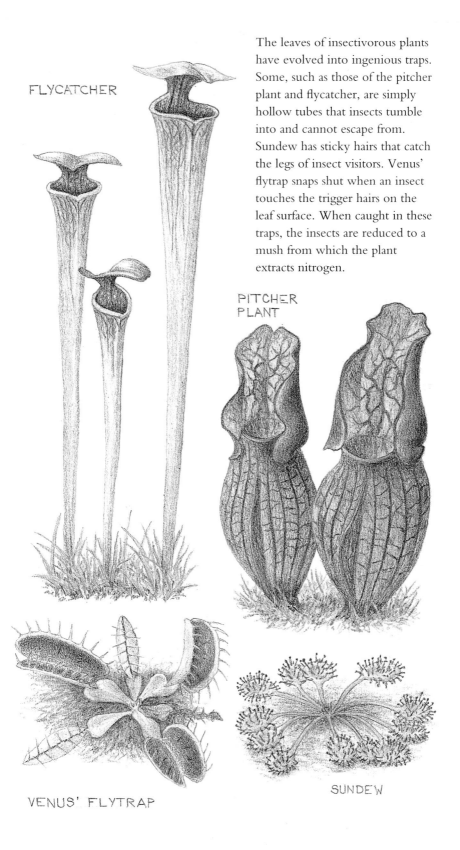

FLYCATCHER

The leaves of insectivorous plants
have evolved into ingenious traps.
Some, such as those of the pitcher
plant and flycatcher, are simply
hollow tubes that insects tumble
into and cannot escape from.
Sundew has sticky hairs that catch
the legs of insect visitors. Venus'
flytrap snaps shut when an insect
touches the trigger hairs on the
leaf surface. When caught in these
traps, the insects are reduced to a
mush from which the plant
extracts nitrogen.

PITCHER
PLANT

VENUS' FLYTRAP

SUNDEW

Plants can spend a lot of energy competing for space to live in and for resources to keep them alive. A better tactic is for species to use their special, individual features to exploit nooks in the larger habitat—places out of bounds to less-well-adapted plants.

These features include:

🌿 Structures that help some aquatic plants float in less crowded, open waters.

🌿 Pigments that enable red algae to live in deep water, using the available dim, blue-green light.

🌿 Mucilage to help seaweeds occupy inhospitable coastal rocks but not dry out at low tide.

🌿 Methods to supply roots with oxygen while growing in oxygen-short places, avoided by other types of plants.

🌿 Streamlined, flexible bodies and strong anchors with which a few hardy plants live in turbulent waters.

🌿 Glands to rid salt marsh plants of excess salts that would kill other species.

🌿 The leaves of a select group of plants that catch and digest insects. These make up for nitrogen missing from bog soils where other types of plants have a hard time growing.

The closer any of us looks at the plant kingdom, the more we are led to question whether an environment exists where plants are not prepared to meet its challenges. And, for the inquisitive mind, in every corner of the world new wonders of plant adaptation are yet to be discovered.

ANCIENT PLANTS

Strange-looking plants inhabited the earth 300 million years ago, before dinosaurs existed. These ancient plants are known only from fossil remains. From their decayed bodies, coal, oil, and natural gas were formed. When the ancient species became extinct, other types of plants slowly evolved to replace them.

CONCLUSION

Ecologists think of the earth as divided into many different plant communities. Among these are the habitats described in this book. Each habitat is recognized for its physical features: for example, the amount of rainfall and wind it receives, its soil conditions, and seasonal changes in temperature. In an aquatic habitat, other features include water movement (is it a still pond or a flowing stream?), saltiness of the water, and the presence of tides.

The *environment* of a habitat is the sum of all such physical features. When describing an environment, scientists may also take into account the effect of the vegetation (especially the dominant species) on the area, and the roles played by both harmful and beneficial animals, fungi, and bacteria.

With the passage of time, some of these physical and biological features change. And so the habitat changes, as do the species of plants and animals that live there. After sensing life-threatening differences in their environment, many animals seek out more favorable locations. But because individual plants cannot move, they depend on the slow, haphazard migration of their species by seed and spore dispersal.

Over millions of years, nature has created a vast array of plants—about 400,000 species living today. Each species is equipped to meet the challenges of a different environment. So while environments change and unprepared plants die out, other better-adapted species take their place in newly composed habitats.

The key to the success of such a system is species diversity. With many, varied choices to work with, nature can keep pace with a constantly changing world and heal the vegetation in damaged places.

In addition, when environmental changes take place slowly, over thousands or millions of years, new species have a chance to evolve—among them, plants with new adaptations for surviving in altered habitats.

During most of human history, our species had only a small impact on the earth and its plant communities. But about 100 years ago the human population began to grow at an alarming rate—doubling, then doubling again in that short period.

In some parts of the world, especially in the tropics, major environmental changes are taking place where the natural vegetation has been cleared to make room for people to live and to grow food. And while supplying the needs of increased numbers of people worldwide, industries have polluted the atmosphere, soil, oceans, and streams with dangerous chemicals. Changes in the earth's climate are even predicted as a result of some human activity.

Because all such changes are happening too fast for most plants to adapt to them, many plant species are becoming extinct. With each lost species, nature's method of healing, which depends on diversity, is being threatened. One of the most important responsibilities that people face today is the pressing need to preserve that diversity in both the plant and animal kingdoms.

During their long occupation of our planet, plants have had plenty of practice in coping with environmental crises. So the changes humans are making do not leave plants completely unprepared.

For example, scientists warn of a thinning ozone layer in the stratosphere, high above the earth. The result: living things will absorb increased amounts of harmful ultraviolet light. But plants on high mountains already adapted to such a problem with red pigments and hairy shoots—a secret of survival shared by species in other places.

Ever since humans began to burn fossil fuels (coal, oil, natural gas) as a source of energy for industry and everyday living, carbon dioxide (CO_2) levels in the atmosphere have been increasing. This gas, in higher than the current concentrations, is poisonous to most animals. But plants thrive on it, converting it into foods during photosynthesis.

Many scientists think that increased CO_2 in the atmosphere will trap heat near the surface of the earth and so increase its temperature. They call it the greenhouse effect. Many cold-weather plants would die as a consequence. But the tropics and deserts are full of species that adapted to heat long ago and are ready to extend their range to new areas if the opportunity arises.

The greenhouse effect may also promote melting of the vast masses of polar ice. With so much water added to the oceans, their level would rise, flooding the land far inland from the present shorelines. Although countless plants would die in the salty water, nature is already prepared to replace them with salt-tolerant seaweeds and salt-marsh species.

Human ingenuity and the remarkable structure of the human body have well prepared our species to survive environmental change. On the other hand, the future of the plant kingdom depends more on the present diversity of species and the plants' amazing adaptations to every conceivable environment. After looking closely at some of their methods, you might say that plants have perfected survival into a fine art.

KEY WORDS

Annual. A plant that germinates from a seed, then grows, flowers, produces seed, and dies in less than a year.

Dominant species. A plant whose leaf canopy, branches, and roots are so extensive that it modifies the local environment.

Ecology. The study of relationships between living things and the environment.

Environment. The sum of the physical and biological features of a habitat, including climate, soil conditions, and the influence of plants and other living things.

Habitat. The place where a plant or animal lives.

Metabolism. The sum of all the chemical reactions occurring in a plant or animal, or in each of its cells.

Perennial. A plant that lives for many years.

Photosynthesis. The chemical process in which light energy is used to form foods from carbon dioxide and water.

Pioneer species. The first plants to occupy an area in a succession.

Species. A group of similar plants or animals, the members of which can breed with each other.

Succession. A change, over time, in an area's plant and animal life.

Symbiosis. A relationship between different species of living things that benefits each member.

Transpiration. The loss of water vapor from plants, mostly from pores in the leaves.

INDEX

NOTE: Page numbers in *italics* refer to illustration captions.